Grades 5-8

Focus On Middle School

Physics

Teacher's Manual

Rebecca W. Keller, PhD

Cover design: David Keller
Opening page: David Keller, Rebecca W. Keller, PhD
Illustrations: Rebecca W. Keller, PhD

Copyright © 2013 Gravitas Publications, Inc.

All rights reserved. No part of this publication may be reproduced, stored in a retrieval system, or transmitted, in any form or by any means, electronic, mechanical, photocopying, recording, or otherwise, without prior written permission from the publisher.

Focus On Middle School Physics Teacher's Manual
ISBN 978-1-936114-67-2

Published by Gravitas Publications, Inc.
www.gravitaspublications.com

Printed in United States

A Note From the Author

This curriculum is designed to give students both solid science information and hands-on experimentation. The middle school material is geared toward fifth through eighth grades, and much of the information in the text is very different from what is taught at this grade level in other textbooks. This is a *real* science textbook, so scientific terms are used throughout. It is not important at this time for students to master the terminology, but it *is* important that they be exposed to the real terms used to describe science.

For students, each chapter has two parts: a reading part in the *Focus On Middle School Physics Student Textbook* and an experimental part in the *Focus On Middle School Physics Laboratory Workbook*. In this teacher's manual, an estimate is given for the time needed to complete each chapter. It is not important that both the reading portion and the experimental portion be concluded in a single sitting. It may be better to have students do these on two separate days, depending on the interest level of the child and the energy level of the teacher. Also, questions not addressed in the *Teacher's Manual* may arise, and extra time may be required to investigate these questions before proceeding with the experimental section.

Each experiment is a *real* science experiment and not just a demonstration. They are designed to engage students in actual scientific investigation. The experiments are simple but are written the way real scientists actually perform experiments in the laboratory. With this foundation, it is my hope that students will eventually begin to think of their own experiments and test their own ideas scientifically.

Enjoy!

Rebecca W. Keller, PhD

How To Use This Manual

Each chapter in this *Focus On Middle School Physics Teacher's Manual* begins by providing additional information for the corresponding chapter in the *Focus On Middle School Physics Student Textbook*. This supplementary material is helpful when questions arise while students are reading the textbook. It is not necessary for students to learn this additional material since most of it is beyond the scope of this level. However, the teacher may find the information helpful when answering questions.

The second part of each chapter in the *Teacher's Manual* provides directions for the experiments in the *Laboratory Workbook* as well as answers to the questions asked in each experiment and review section. All of the experiments have been tested, but it is not unusual for an experiment to produce an unexpected outcome. Usually repeating an experiment helps both student and teacher see what might have occurred during the experimental process. Encourage the student to troubleshoot and investigate all possible outcomes. However, even repeating an experiment may not produce the expected outcome. **Do not worry if an experiment produces a different result.** Scientists don't always get the expected results when doing an experiment. The important thing is for students to learn about the scientific method and to make observations, think about what is taking place, and ask questions.

Getting Started

The experimentation process will be easiest if all the materials needed for the experiment are gathered together and made ready before beginning. It can be helpful to have a small shelf or cupboard or even a plastic bin dedicated to holding most of the necessary chemicals and equipment. The following *Materials at a Glance* chart lists all of the materials needed for each experiment. An additional chart lists the materials by type and quantity. A materials list is also provided at the beginning of each lesson.

Laboratory Safety

Most of these experiments use household items. Extra care should be taken while working with all materials in this series of experiments. The following are some general laboratory precautions that should be applied to the home laboratory:

- Never put things in your mouth without explicit instructions to do so. This means that food items should not be eaten unless tasting or eating is part of the experiment.

- Wear safety glasses while using glass objects or strong chemicals such as bleach.

- Wash hands before and after handling all chemicals.

- Use adult supervision while working with electricity and glassware, and while performing any step requiring a stove.

Materials at a Glance

Experiment 1	Experiment 2	Experiment 3	Experiment 4	Experiment 5
tennis ball yarn or string (3 meters [10 ft]) paper clip marble **Optional** awl, penknife, or other sharp instrument to pierce tennis ball tape 1 or more small objects such as building blocks	Slinky several paper clips 1-2 apples 1-2 lemons or limes 1-2 oranges 1-2 bananas spring balance scale or food scale meter stick, yardstick, or tape measure tape	small to medium size toy car stiff cardboard wooden board (more than 1 meter [3 ft] long) straight pin or tack (1-3) small scale or balance 1 banana, sliced 10 pennies meter stick, yardstick, or tape measure tape	glass marbles of different sizes (several) steel marbles of different sizes (several) cardboard tube, .7-1 meter [2.5-3 ft] long scissors black marking pen ruler letter scale or other small scale or balance	10-20 copper pennies aluminum foil paper towels salt water, 30-45 ml (2-3 Tbsp) salt per 240 ml (1 cup) water voltmeter ** 2 plastic-coated copper wires, each 10-15 cm (4-6 in) long duct tape (or other strong tape) scissors wire cutters steel wool **Optional** vinegar (240 ml [1 cup])

Experiment 6	Experiment 7	Experiment 8	Experiment 9	Experiment 10
small glass jar with lid aluminum foil paper clip duct tape (or other strong tape) plastic or rubber rod (or balloon) silk fabric (or use hair) scissors ruler **Optional** several dry cell batteries of different sizes and shapes for observation	1.2 meters (4 ft) insulated electrical wire 6v or larger (up to 12v) battery insulating materials (e.g., styrofoam, plastic, cloth) small light bulb (Radio Shack flashlight lamp #272-1163 with a rated voltage of 6v or comparable bulb) electrical tape several small resistors scissors wire cutters **Optional** 2 alligator clips	metal rod (a large nail [such as an 8.9 cm [3½"] long 16d flathead nail], a screwdriver, or other) electrical wire 10-20 paper clips 6v-12v battery electrical tape scissors wire cutters magnets, bar or horseshoe (2 or more) **Optional** thin magnet that can be cut iron filings 60 ml (1/4 cup) corn syrup shallow dish iron nail alligator clips (2)	2 prisms (glass or plastic) flashlight metal can, open at both ends aluminum foil rubber band laser pointer long wooden craft stick colored pencils duct tape (or other strong tape) **Optional** bowl filled with water styrofoam or other material that floats eyedropper	student-selected materials

** An inexpensive voltmeter can be purchased at any store that supplies electrical equipment. Make sure the voltage scale is low enough to detect small voltages—about 0.5v.

Materials at a Glance
By type with total quantities

Equipment	Materials	Materials (cont.)	Food Items
battery, 6v or larger (up to 12v) flashlight laser pointer light bulb, small (Radio Shack flashlight lamp #272-1163 with a rated voltage of 6v or comparable bulb) prisms, glass or plastic, 2 scale: spring balance, food, or letter scissors voltmeter (inexpensive, from electrical equipment store, voltage scale should read as low as .5v) wire cutters **Optional** alligator clips (2) awl, penknife, or other sharp instrument to pierce a tennis ball	*Focus On Middle School Physics Laboratory Workbook* aluminum foil board, wooden (more than 1 meter [3 ft] long) can, metal, open at both ends cardboard, stiff, 1 piece cardboard tube, .7-1 meter (2.5-3 ft) long fabric, silk (or hair) insulating materials (e.g., styrofoam, plastic, cloth) Jar, small glass with lid magnets, bar or horseshoe (2 or more) marbles, glass, several different sizes marbles, steel, several different sizes meter stick, yardstick, or tape measure misc. materials, student selected (Exper. 10) paper clips, 10-20 paper towels, several pen, marking, black pencils, colored pennies, 10-20 pin, straight, or tack (1-3)	resistors, small, several rod, plastic or rubber (or balloon) rod, metal (a large nail [such as an 8.9 cm [3½"] long 16d flathead nail] or a screwdriver can be used for the rod) rubber band ruler sticks, craft, long wooden Slinky steel wool tape (any) tape, duct tape, electrical tennis ball toy car, small to medium size wire, plastic-coated copper, 2.5 meters (8 ft) yarn or string, 3 meters (10 ft) **Optional** batteries, dry cell, different sizes and shapes for observation (several) dish, shallow eyedropper iron filings magnet, thin (to cut up) nail, iron objects, small, 1 or more (such as building blocks) styrofoam or other material that floats	1-2 apples 1-2 lemons or limes 1-2 oranges 1-3 bananas salt water, 30-45 ml (2-3 Tbsp) salt per 240 ml (1 cup) water **Optional** 60 ml (1/4 cup) corn syrup vinegar water in bowl

Contents

CHAPTER 1:	**WHAT IS PHYSICS?**	**1**
	Experiment 1: It's the Law!	7
	Review	11
CHAPTER 2:	**FORCE, ENERGY, AND WORK**	**12**
	Experiment 2: Fruit Works?	18
	Review	22
CHAPTER 3:	**POTENTIAL AND KINETIC ENERGY**	**23**
	Experiment 3: Smashed Banana	28
	Review	33
CHAPTER 4:	**MOTION**	**34**
	Experiment 4: Moving Marbles	39
	Review	44
CHAPTER 5:	**ENERGY OF ATOMS AND MOLECULES**	**45**
	Experiment 5: Power Pennies	49
	Review	54
CHAPTER 6:	**ELECTRICAL ENERGY AND CHARGE**	**55**
	Experiment 6: Charge It!	58
	Review	62
CHAPTER 7:	**MOVING ELECTRIC CHARGES AND HEAT**	**63**
	Experiment 7: Let It Flow	67
	Review	71
CHAPTER 8:	**MAGNETS AND ELECTROMAGNETS**	**72**
	Experiment 8: Wrap It Up!	76
	Review	81
CHAPTER 9:	**LIGHT AND SOUND**	**82**
	Experiment 9: Bending Light and Circle Sounds	87
	Review	93
CHAPTER 10:	**CONSERVATION OF ENERGY**	**94**
	Experiment 10: On Your Own	98
	Review	103

Chapter 1: What Is Physics?

Overall Objectives	2
1.1 Introduction	2
1.2 The Basic Laws of Physics	2
1.3 How We Get Laws	2
1.4 The Scientific Method	3
1.5 Summary	6
Experiment 1: It's the Law!	7
Review	11

Time Required

Text reading 30 minutes
Experimental 1 hour

Materials

tennis ball
yarn or string (3 meters [10 ft])
paper clip
marble
Optional: awl, penknife, or other sharp instrument to pierce the tennis ball
tape
one or more small objects such as building blocks

Overall Objectives

This chapter will introduce students to a fundamental concept in physics called physical laws. The students will also examine the scientific method.

1.1 Introduction

In this section the students begin their inquiry into physics by making observations about the physical world. Begin a discussion by asking the students to describe several observations they have made.

For example:

- *What happens when you put on the brakes while riding a bicycle? Do the tires stop immediately? Do they skid?*
- *What happens when you throw a ball into the air? Does it reach the clouds? Does it come down in the same spot?*
- *What happens when you turn on a flashlight? How far can you see the light? Can you see the beam from a flashlight in the daytime?*

Encourage the students to discuss as many observations as they can think of. There are no "right" answers, and at this point, it is not important to know the reasons why something happens.

1.2 The Basic Laws of Physics

Ask the students what a law is, such as a law against driving too fast or a law against stealing. Ask if these laws are ever broken and, if so, why they are.

Ask the students some questions about what they have consistently observed in the physical world. For example:

- *Have you ever thrown a ball and have it not come down (except when it gets stuck somewhere like in a tree)?*
- *Does ice always float?*
- *Does the Sun always come up in the morning?*

Explain that laws in physics differ from the kinds of laws that govern our country. In physics a law is an overall principle or relationship that remains the same and is not broken.

1.3 How We Get Laws

Have a discussion with the students about how we make laws for our country, city, or state. Discuss how the making of city, state, or federal laws involves a long process where several people decide what kinds of laws to make. Because there are different people making the laws, some laws are different from city to city or from state to state. For example, the speed limit is different in different states because the governments of the states have different ideas about what the speed limit should be. Explain that governmental laws are laws we make ourselves and, because of this, the laws sometimes differ.

Ask the students if they think physical laws are laws we make ourselves. Do physical laws differ from state to state or country to country? Do they think that a baseball hit from a ballpark in Alaska or Hawaii might be able to reach the clouds? Will ice float in Arizona, but sink in New Jersey? The answer is "no," a ball will not reach the clouds in Alaska or Hawaii, and "yes," ice still floats if it's in New Jersey.

Explain to the students that physical laws are not laws we make up ourselves. They are regularities that scientists have discovered in the way things behave. The physical world is ordered, reliable and consistent. This orderliness means there are underlying physical laws, or general principles, that we can discover to better understand the world.

Explain to the students that physical laws are described by mathematics. Because the universe is ordered, mathematics can be used to precisely describe the laws that govern it.

1.4 The Scientific Method

Although scientific investigation began with Aristotle over two thousand years ago, the foundations of the scientific method were not established until the 13th century by Roger Bacon and further elucidated in the 17th century by Rene Descartes.

The scientific method has 5 steps:

1. *observation*
2. *formulating a hypothesis*
3. *experimentation*
4. *collecting results*
5. *drawing conclusions*

The first step in the scientific method is *observation*. Have a discussion with the students about how observations are made. Give some examples of observations that use sight, hearing, taste, smell, or touch. For example:

- *Salt is poured on icy roads when it snows.*
- *Lemons are sour. Oranges are sweet.*
- *The sky is blue.*
- *Ice cubes float in soda, water, and milk.*
- *Thunder is loud when lightning is close.*
- *Steel balls or marbles sometimes feel cold, but cotton balls do not.*

Have the students turn these observations into questions:

- *Why is salt poured on icy roads when it snows?*
- *Why are lemons sour but oranges sweet?*
- *Why is the sky blue?*
- *Why do ice cubes float in soda, water, and milk? Will ice cubes float in oil?*
- *Why is thunder loudest when lightning is closest?*
- *Why do steel balls or marbles sometimes feel cold, but cotton balls do not?*

Explain that after making observations and asking questions about those observations, the next step in the scientific method is *formulating a hypothesis*. Hypotheses are guesses. That is, from observations and the questions about those observations, a statement can be made about why something is or behaves in a certain way. Although a scientist attempts to make a good guess, the hypothesis may prove to be incorrect. For example:

- *Salt is put on roads to make rubber tires sticky.*
  (Salt is actually used to lower the freezing temperature of ice, causing the ice to melt.)
- *Lemons are sour because they have no sugar.*
  (Lemons have some sugar, just less sugar than oranges.)
- *Oranges are sweet because they have sugar.*
- *The sky is blue because all of the other colors get absorbed by water in the atmosphere.*
  (The sky is actually blue because of light scattering, called Raleigh scattering.)
- *Ice cubes float because they repel soda, water, and milk.*
  (Ice cubes float because ice is less dense than liquid water.)
- *Ice cubes will not float in oil.*
  (It depends on the density of the oil being used.)
- *Thunder is louder when it is closer because the sound hits our ears sooner and has less chance to go someplace else if we are close.*
  (Thunder travels as a sound wave, and the farther we are from the thunder, the more the sound gets dampened as it hits molecules in the air while it is traveling.)
- *Steel balls and marbles sometimes feel cold in our hands because they allow heat to exchange but cotton balls do not allow heat to exchange.*

These are some examples of hypotheses. Explain to the students that not every hypothesis is correct, and when scientists formulate a hypothesis, they do not already know the correct answers. A scientist is making an educated guess.

The next step in the scientific method, *designing an experiment,* will help determine whether or not the hypothesis is correct. Explain that because a hypothesis is a guess, a scientist must do something, like design an experiment, to test whether or not the hypothesis is correct.

Using the example in this section of the textbook, have a discussion with the students about the way in which an experiment can be designed to test the boy's hypothesis that salt makes rubber sticky. Or, help the students think of ways they might test for sugar in a lemon or test some other hypothesis. For example:

- *Lemon juice could be collected and the water evaporated. Sugar might be visible in the remaining residue.*

- *Lemon juice residue could be compared to orange juice residue.*

- *Ice cubes could be placed in several different liquids, such as water, milk, soda, and oil, to determine if ice cubes float in these liquids.*

Tell the students that sometimes it is difficult to design experiments and not every question can be answered. For example, it may be difficult for a student to design an experiment to test conclusively whether or not lemons have sugar. The residue may contain other chemicals that do not allow one to say with certainty whether or not a lemon has sugar. Also explain that there is not necessarily one right way to do an experiment although there are certain procedures that, when included, will usually make the experiment better. For example, controls are used to make sure the experimental setup is working properly. Controls can be either positive or negative.

A positive control tells the scientist what the results of an experiment might look like if the hypothesis is true. For example, if lemon juice contains sugar, and if it is possible to evaporate the water from the lemon juice leaving the sugar behind, then a scientist might want to know what sugar would look like when it has been separated from water by evaporation. To find out, a positive control made only of sugar and water could be used. The scientist would mix sugar and water together, let the water evaporate, and then examine the residue. This control could then be compared with the experimental results to find out if the residues looked similar. If they did, the conclusion might be that there is sugar in lemon juice.

To discover what the results of an experiment should not look like, a scientist can use a negative control. Also, if a positive control might be confusing, it can be helpful to use a negative control. For example, it may not be easy to tell the difference between salt water and sugar water. A scientist might set up a negative control where salt is used instead of sugar. If evaporated salt water looks similar to evaporated sugar water, then it won't be easy to tell if the residue from the lemon juice is sugar or salt. Some other test is needed, such as tasting the residue.

Once an experiment has been designed, *results are collected* as the experiment is carried out. This is the next step in the scientific method. Explain to the students that it is very important that *all* of the results be recorded. This includes results that were not expected. Sometimes major scientific discoveries happen from scientists getting results that were not at all expected. A good scientist has a keen sense of observation and does not let expectations of what will happen determine what gets recorded. Some possible results for lemon juice might be:

- *No residue was found after the water evaporated.*
- *Residue was found.*
- *The residue did not taste like anything.*
- *The residue tasted salty (or sweet or sour).*

The final step in the scientific method is *drawing a conclusion.* In a conclusion the scientist evaluates the results of the experiment and tries to make a statement regarding the hypothesis. For example, if residue was found in lemon juice and the residue tasted sweet, the scientist can conclude that the hypothesis that "lemons have no sugar" may not be correct. At this point, the scientist needs to determine how conclusive the data are and check the reliability of the experimental set up. It is important that the conclusions be valid and not state something that the data haven't shown.

1.5 Summary

Go over the summary statements with the students. Discuss any questions they might have.

Experiment 1: It's the Law!   Date: _____

Objective In this experiment we will use the scientific method to determine Newton's First Law of Motion.

Hypothesis _____

_____

Materials

    tennis ball
    yarn or string (3 meters [10 ft])
    paper clip
    marble

Experiment

PART I

❶ Take the tennis ball outside, and throw it as far as you can. Observe how the ball travels through the air. In the space below, sketch the path of the ball.

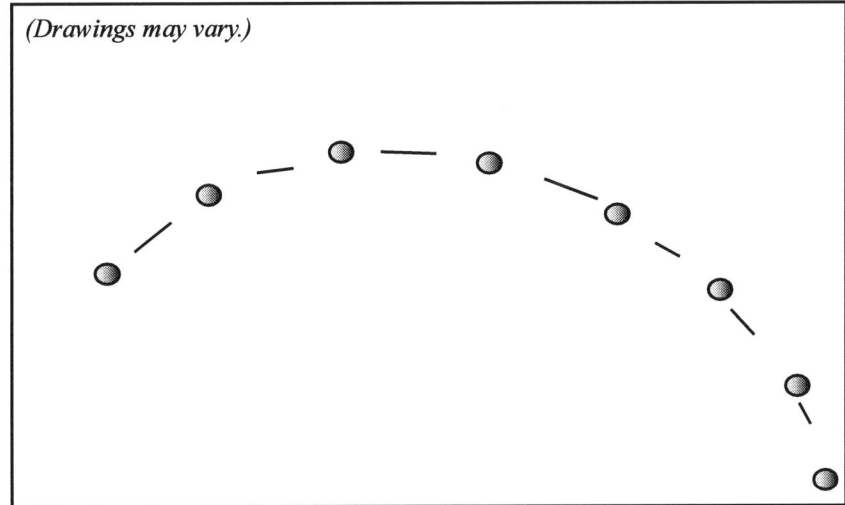

(Drawings may vary.)

In this experiment the students will discover Newton's First Law of Motion by observing the motion of a tennis ball and a marble.

Newton's First Law of Motion is also called the Law of Inertia. The students will look more carefully at motion and inertia in Chapter 4. However, in this experiment the objective is to show the students that, by observation, they can discover physical laws.

Newton's First Law of Motion can be stated as:

> *A body will remain at rest or in motion until it is acted on by an outside force.*

In the first part of this experiment, the students are to observe how a ball flies through the air. They should notice that the ball will go up and come down in some kind of arc every time they throw it. The arc can be shallow or sharp depending on how they throw the ball.

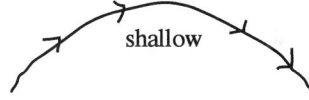
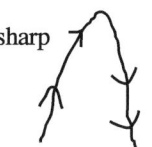

Challenge them to throw the ball so that it won't come down.

Ask them if they can get the ball to go up and down in a different pattern, such as:

By attaching a tennis ball to one end of a long string, the students will be able to observe a difference in how the ball will travel once it is thrown.

The method shown is only one way to attach a string to a tennis ball. Several other methods were tried, but it seems that the paper clip works the best. It is somewhat difficult to puncture the tennis ball with the paperclip, so supervise the students. It might help to put a small hole in the tennis ball with a penknife, ice-pick, or awl before inserting the paper clip.

If you do not want to puncture the tennis ball, the string can be wrapped several times around the ball and secured with tape. A longer piece of string should be used for this.

❷ Now, take the string or yarn and, using the paper clip, attach it to the tennis ball. To do this, open the paper clip up on one side and curve the end as follows:

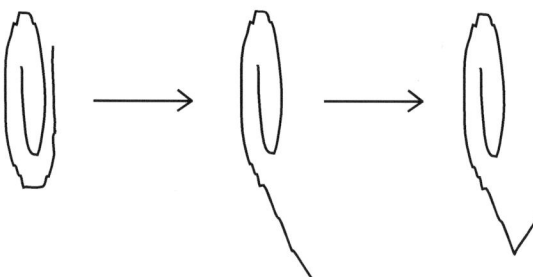

❸ Put the extended curved end of the paper clip into the tennis ball by gently pushing and twisting.

❹ Next, tie the string to the end of the paper clip.

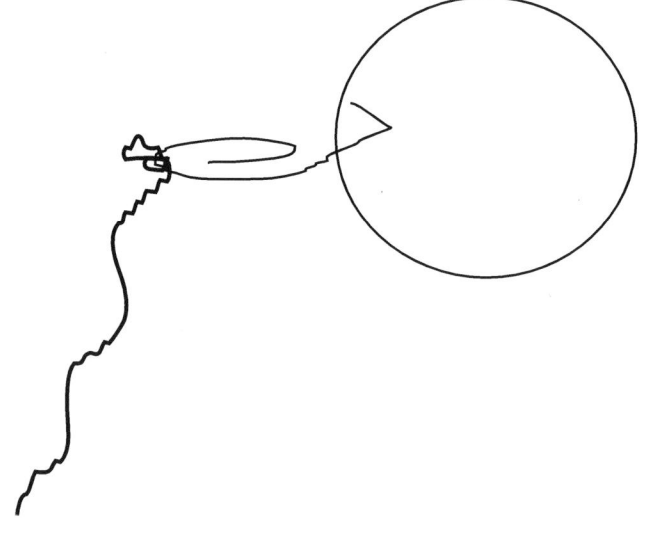

❺ Holding onto one end of the string, again throw the ball into the air as far as you can. Note how the ball travels and, in the space below, record what you see. Do this several times.

(Drawings may vary.)

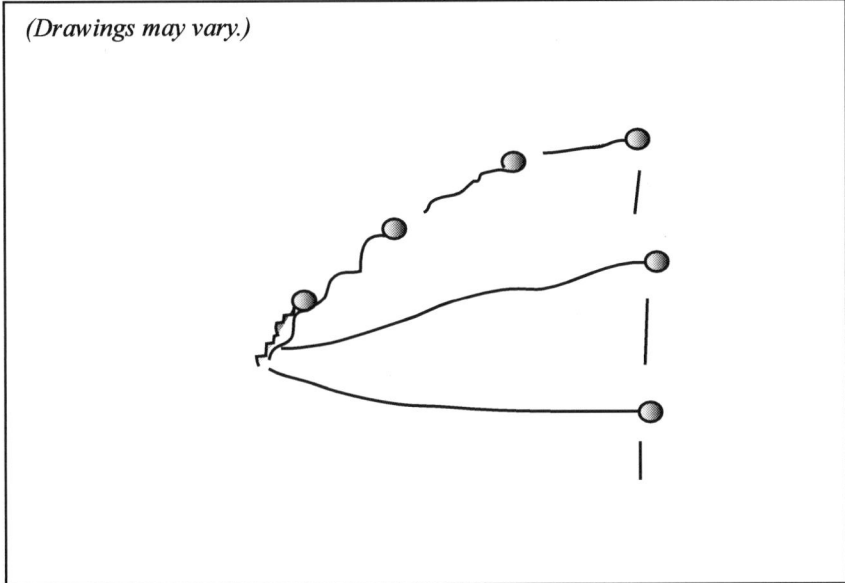

Part II

❶ Take the marble and find a straight, clear path on a smooth area of the floor or outdoors. Roll the marble, and record how it travels. Note where and how it stops or changes direction. Do this several times, and record your observations in the next box.

The trajectory of the tennis ball will now be different. When the students throw the ball, it will begin similarly, but when the string has reached its full length, the ball will abruptly stop and fall to the ground.

Have the students throw the ball several times. Ask them if they can change how the ball falls to the ground. They should notice that if they shorten the string, the ball does not travel as far as when the string is longer. They should also notice that if they do not throw the ball very far and it does not reach the end of the string, the ball will travel almost as if there were no string attached to it.

Without telling them the answer, help the students see that the ball's trajectory only changes when the string can act on it.

In Part II, the students will examine how a marble rolls. Have them roll the marble on a smooth surface. They should notice the marble traveling mostly straight.

Have the students compare the marble traveling on a smooth surface with one rolling on a rough surface. Discuss with the students why they think the marble may travel differently.

You can also have the students place obstacles, such as small building blocks, in front of the marble. They should be able to observe the marble traveling straight on a smooth surface until it contacts an obstacle.

Help the students see that the marble's trajectory is not changed unless it is contacted by something—like a rough surface or a building block.

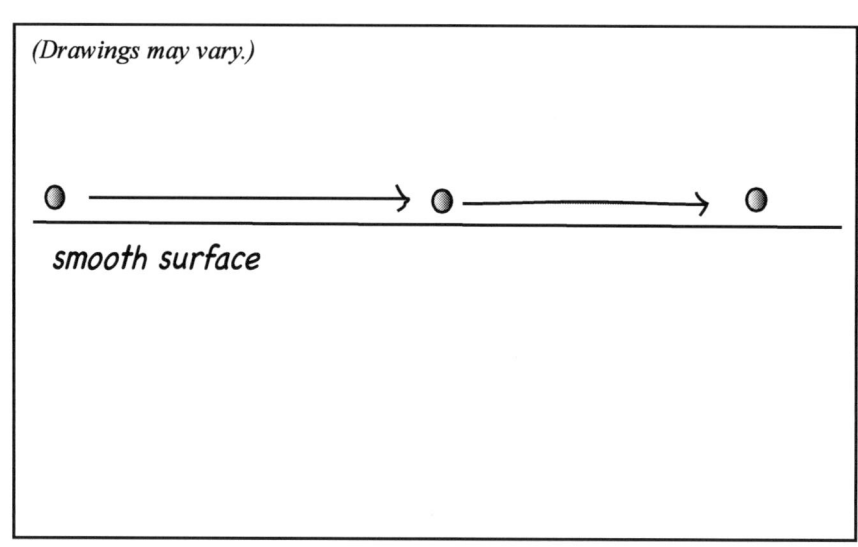

❷ Repeat Step 1 using a rough surface on which to roll the marble.

Help the students draw conclusions based on the data they have collected.

Some possible conclusions are:

- The tennis ball goes up and always comes down.
- The string keeps the tennis ball from going all the way up because it pulls the tennis ball back.
- The marble travels on the smooth surface in a straight line, but the rough surface keeps the marble from traveling straight.
- The marble on a smooth surface changes direction only when it hits a block.

After the students have thought about their data and drawn some conclusions, discuss Newton's First Law of Motion *(A body will remain at rest or in motion until it is acted on by an outside force)*. Show them how they were able to observe the same things that Newton observed and that they too could discover a fundamental law of physics.

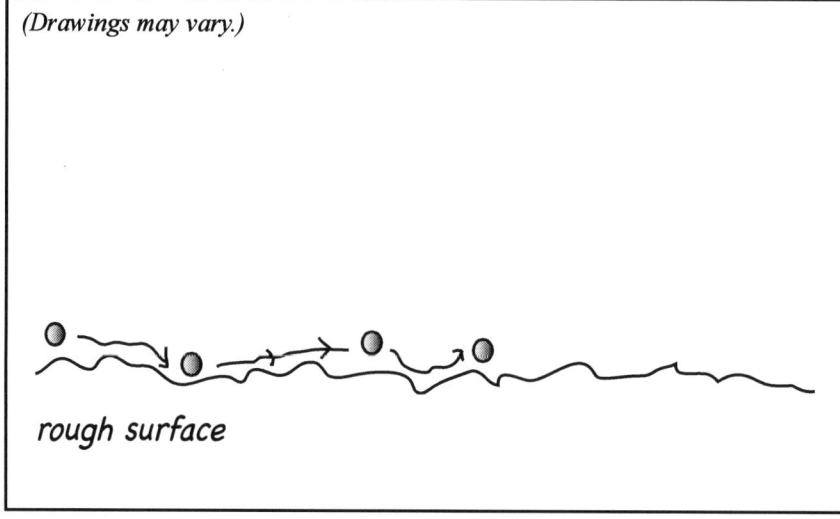

Conclusions

Draw some conclusions about your results and record them below.

_____
_____
_____
_____
_____
_____
_____
_____
_____
_____

Review

Define the following:

physics — *the study of how things move and behave in nature*

physical law — *a precise statement about how things behave in the physical world*

List the 5 steps of the scientific method

1. *observation*
2. *forming a hypothesis*
3. *experimentation*
4. *collecting results*
5. *drawing conclusions*

Chapter 2: Force, Energy, and Work

Overall Objectives	13
2.1 Introduction	13
2.2 Force	13
2.3 Balanced Forces	15
2.4 Unbalanced Forces	15
2.5 Work	16
2.6 Energy	16
2.7 Summary	17
Experiment 2: Fruit Works?	18
Review	22

Time Required

    Text reading   30 minutes
    Experimental   1 hour

Materials

    Slinky
    several paper clips
    1-2 apples
    1-2 lemons or limes
    1-2 oranges
    1-2 bananas
    spring balance scale or food scale
    meter stick, yardstick, or tape measure
    tape

Overall Objectives:

This chapter will introduce the students to the fundamental concepts of force, energy, and work. These concepts can be difficult to understand; however, it is not important that the students completely grasp everything about these concepts. This chapter is only a qualitative introduction. The students should be encouraged to think about the concepts but not necessarily to understand all of their subtleties.

2.1 Introduction

The terms energy, work, and force are introduced in this section. Have a discussion with the students concerning their own ideas about these terms.

Ask them:

- *What is energy?*
- *Can energy be created or destroyed?*
- *What happens to the energy in a battery when the battery dies?*
- *What is work?*
- *When you move bricks from the front yard to the back yard—is that work?*
- *Is lifting a book work?*
- *Is dropping a book work?*
- *What is force?*
- *Can you give some examples of force?*

Most of their answers may not be correct, and they may have some misconceptions about energy, work, and force. For example, it is a common misconception that the energy in a battery is destroyed when the battery dies. Students will better understand the nature of energy by examining this misconception. In reality, the usable energy has only been converted into a form that cannot be used any longer.

2.2 Force

This section examines the nature of force in more detail. By definition, a force is something that:

- *changes the position, shape, or speed of an object.*

Using this definition, lead a discussion about some of the forces the students experience every day. Following are some examples of questions that could be asked to begin the discussion:

- *What happens when you jump?*
- *What happens when you collide with another player on the football field?*
- *What happens when you lift a heavy object?*
- *What happens when you drop a heavy object?*
- *What happens when you pull or push on a door?*
- *What happens to a marshmallow when you squeeze it?*
- *What happens to a steel ball when you squeeze it?*

Explain that in all of these examples forces are acting.

Discuss gravitational force, which is an attraction between any two bodies that have mass. Mass is the property that makes matter resist being moved. Weight is a force caused by the Earth's gravity.

Mass has two very important roles in physics. First, any two bodies with mass will attract each other by gravitation. So, the amount of mass determines how strong gravitational forces are. Big objects with large masses attract each other much more than do small objects with small masses. The weight of any object is the gravitational attraction between the object and the Earth (as measured at the Earth's surface). So weight is a force, not a mass.

Second, the mass of an object controls its inertia—the tendency of any body to remain in motion unless acted on by a force or to stay at rest unless acted on by a force. A body with a large mass will accelerate slowly when acted on by a force, and a body with a small mass will accelerate quickly when acted on by the same force. At first glance, this seems like a completely different property than gravitational attraction, and it is indeed a very strange fact that one property of matter, mass, controls two seemingly different things. The fact that gravitational mass and inertial mass are the same is called the equivalence principle, which is the basis for *Einstein's theory of gravitation* (also called general relativity). Both inertia and gravitation arise from one thing—the curvature of space itself!

The students will look at inertial mass in more detail in Chapter 4.

The equation for gravitational force is:

$$F = G * m_1 m_2 / d^2$$

where G is the universal gravitational constant, m_1 is the mass of the first body, m_2 is the mass of the second body, and d is the distance between the two bodies.

We can see from the equation that a body that has more mass will have a greater force. Explain to the students that because their own body mass is so much less than the mass of the Earth, their gravitational force is also much smaller, so they cannot pull the Earth toward them. This is why they come back to the Earth when they jump, rather than the Earth shifting upwards to meet them. (Do they think it might be possible to shift the Earth by having everyone jump at the same time? Why or why not?)

2.3 Balanced Forces

Explain the concept that when objects aren't moving, forces are balanced.

The most fundamental equation defining force is *Newton's Second Law of Motion:*

$$F = ma \text{ or } a = F/m$$

where *F* is force, *m* is mass and *a* is acceleration. This equation says that the force exhibited by an object is equal to its mass times its acceleration (see Section 2.4). This equation shows that if the acceleration of an object is zero, then the net force acting on that object is also zero. Or conversely, if a given force acts on an object causing it to accelerate, an object having a small mass will accelerate more than an object with a large mass.

Have the students look at the illustration in this section of the textbook. Have them take note of the ball shown sitting on a shelf. In the diagram there are arrows pointing in opposite directions showing that the ball is pushing down on the shelf and at the same time the shelf is pushing up on the ball. Point out that the forces are equal but acting in opposite directions.

In this case, the net force is zero, and therefore the ball does not move.

Fnet = [force of the ball pushing down] - [force of the shelf pushing up] = zero

The forces are balanced: i.e., they cancel each other out.

Explain that objects that are moving but not accelerating also have balanced forces. Discuss the diagram in the textbook that shows an air hockey puck and point out that, although the hockey puck is moving, it is not accelerating.

2.4 Unbalanced Forces

When forces become unbalanced, an object will accelerate. Acceleration is the change in an object's speed over time, such as a car speeding up from a stoplight or a ball speeding up when it is dropped. An object's speed also changes when it slows down, so slowing down is also a form of acceleration. Just as a massive object is hard to speed up, it is also hard to slow it down.

Have the students try to come up with some examples of unbalanced forces. For instance:

- *Does a ball thrown in the air have balanced or unbalanced forces? Why?*
  (unbalanced because it speeds up and slows down)

- *Does an airplane when it takes off have balanced or unbalanced forces?*
  (unbalanced—speeds up to take off)
- *Does a car starting up from a stoplight have balanced or unbalanced forces?*
  (unbalanced)

2.5 Work

The concept of work may be difficult to understand because, when we hear the word "work," we think of mowing the lawn or doing the laundry. However, in physics, work is defined as:

$$\text{work} = \text{distance} \times \text{force}$$

The illustration in the student textbook shows that, for the same amount of force, the work a short weight lifter does is less than the work a tall weight lifter does because the distance of the lift is less for the short weight lifter.

Have a discussion about other examples relating work, distance, and force. For example:

- *If you carry a box of books up one flight of stairs, and your brother carries the same box up two flights of stairs, who has done more work?*
  (your brother)

  How much more work has he done?
  (Exactly twice the amount of work was done.)

- *If you carry a box of books up one flight of stairs, and your brother carries a box of books that has half the mass up the same flight of stairs, who has done more work?*
  (you have)

  How much more?
  (exactly twice as much)

Help the students think of some of their own examples. The more examples they discuss, the more they will understand the relationship between work, distance, and force.

2.6 Energy

Energy is another concept that can be difficult to grasp because when we hear the word "energy," many different ideas come to mind. Basically, *energy* gives objects the ability to do work. The different kinds of energy, such as potential energy, kinetic energy, and heat energy, will be discussed in more detail in later chapters. At this point the important thing is for students to begin thinking about where the ability for an object to do work comes from.

Ask the students to list some sources of energy that can provide an object with the ability to do work.

- *What do you need to make a car run?*
  (gasoline)

- *What do you need for a flashlight to work?*
  (batteries)

- *What do you need for a CD player to work?*
  (plug into an outlet—electricity)

- *What do you need to have to carry books up a flight of stairs?*
  (muscles, food)

2.7 Summary

Go over the summary statements with the students. Discuss any questions they might have.

FOCUS ON MIDDLE SCHOOL PHYSICS
Teacher's Manual

In this experiment the students will try to determine how much work a variety of fruits can do. Remind the students that:

work = distance x force

Have the students read the entire experiment and then help them think of a possible objective. For example:

- *Using a Slinky, we will find out if a banana can do more work than an orange.*
- *We will measure the work that fruit can do.*
- *We will find out if two bananas do more work than one.*

Have the students make a guess about which fruit will do more work. They should be able to tell which one is the heaviest just by feeling the weights of the different fruits in their hands. Have them state their theory in the hypothesis. For example:

- *A banana is heavier than a lemon and will do more work.*
- *The orange is lighter than the apple and will do less work.*
- *Two bananas will do more work than one banana because two bananas weigh more.*

Using a food balance or a small scale, have the students weigh each piece of fruit and record their weights in the chart.

Experiment 2: Fruit Works?   Date: _____

Read through all the steps of this experiment. Then write an objective and a hypothesis.

Objective _____

_____

Hypothesis _____

_____

Materials

Slinky
several paper clips
1-2 apples
1-2 lemons or limes
1-2 oranges
1-2 bananas
spring balance scale or food scale
meter stick (or yardstick) or tape measure
tape

Experiment

❶ Try to determine, just by "weighing" each piece of fruit in your hands, which piece will do the most work and which piece will do the least work on the spring that is in the scale.

❷ State your prediction as the hypothesis.

❸ Next, weigh each piece of fruit on the balance or food scale.

❹ Record the weights in the following chart.

Fruit	Weight (grams or ounces)

5. Prepare the fruit for the experiment. Take a paper clip and stretch one side out to make a small hook like you did in Experiment 1. Place the hook in one of the pieces of fruit.

   Repeat for each different kind of fruit you will be testing.

6. Next, take the Slinky and hold it up to the level of your chest. Allow 10 to 15 coils to hang below your hand. You will have to hold most of the Slinky in your hand.

7. Measure the distance from the floor to the bottom of the Slinky with the meter stick, yardstick, or tape measure. Record your result below.

Distance from floor to Slinky with no fruit attached

8. Take a piece of fruit with a hook in it and attach it to the end of the Slinky. Hold the Slinky at the same height as in Step 6 with the same number of coils hanging below your hand. Allow the Slinky to be pulled down by the fruit.

9. Use the meter stick, yardstick, or tape measure to measure from the end of the Slinky to the floor. Record your results in the following chart in the *Distance Floor to Slinky (With Fruit)* column.

10. Repeat Steps 8 and 9 with different kinds of fruit. Record your results each time.

Have the students use the paper clips to create hooks to use to attach the fruit to the Slinky. We found that the paper clips worked fairly well, but the younger kids found tape more effective. The fruit can be fixed to the Slinky in any manner. You might ask the students to come up with their own ideas for attaching the fruit.

Students will have to experiment with the Slinky and the number of coils that hang down. We found that it worked fairly well to have a student hold most of the coils in their hand and allow only a few coils to fall below the hand. Also, instead of holding the Slinky, it can be attached to a branch of a tree or a fixed ledge of some sort. Just make sure the Slinky is free to extend and does not contact any other surface and that the Slinky is at the same distance from the floor each time.

Have the students first measure the distance from the end of the Slinky to the floor without a piece of fruit attached to it. The distance should be about .6-1 meter (2 to 3 feet). Make sure that once this distance is measured, the number of coils allowed to extend is not altered. If the distance is too short (that is, the fruit ends up touching the ground), reduce the number of coils used and re-measure the distance to the floor before and after attaching the fruit. You may want to have them test the heaviest fruit first.

Have the students attach a piece of fruit to the last coil of the Slinky, and allow the coils to extend. Have them measure the distance from the ground to the bottom of the Slinky for each piece of fruit that is tested and then record the distance in the chart in the *Results* section.

Have the students subtract the distance the Slinky extended without any fruit on it from the distance it extended with the fruit on it. This will give the net displacement. Have them record this answer in the column marked *Distance Extended*.

Have the students use the following equation to determine how much work was done by each piece of fruit.

work = distance x force

where *force* is the weight of the fruit.

Explain the measurement system that the students are using:

When using the metric measurement system in the equation above, the unit of measure of work is the kilogram-meter. For example, 2 kilograms x 2 meters = 4 kilogram-meters of work. (A gram is equivalent to .001 kg.)

When using the British measurement system in the equation above, the unit of measure of work is the foot-pound. For example, 2 pounds x 2 feet = 4 foot-pounds of work. (An ounce is equivalent to .0625 lb.)

NOTE:
In this experiment weight and mass are being used interchangeably, even though they are not the same thing. (Mass is discussed in Chapter 4, Section 4.3.)

Technically, when we weigh something we are measuring its mass times the gravitational acceleration (force of gravity). Gravitational acceleration is equal to 1 and is the same everywhere on Earth. For this reason and for the purposes of this experiment we therefore use mass and weight interchangeably.

Results

❶ In each row of the *Distance Floor to Slinky (No Fruit)* column, write the distance you recorded in Step 7. Then subtract this distance from each of the distances you recorded in the *Distance Floor to Slinky (With Fruit)* column. This gives you the distance the Slinky was extended by each piece of fruit.

Fruit	Distance Floor to Slinky (With Fruit)	Distance Floor to Slinky (No Fruit)	Distance Extended

❷ Calculate the work each piece of fruit has done. Record your answers in the following chart.

Fruit	Work

❸ What would happen if you attached two pieces of the same kind of fruit to the Slinky? How much work would be done?

Prediction _____

_____

_____

❹ Test your prediction and calculate the work that was done by the two pieces of fruit. Record your data in the charts below.

2 Pieces of Fruit	Weight	Distance: Floor to Slinky (With Fruit)	Distance Floor to Slinky (No Fruit)	Distance Extended

2 Pieces of Fruit	Work

Conclusions

Draw some conclusions about your results and record them below.

_____

_____

_____

_____

_____

_____

_____

Now have the students predict what would happen if they attached two bananas or two other pieces of the same kind of fruit to the Slinky. They should predict that two pieces of fruit will do more work than one piece of fruit. Have them test this prediction by attaching two pieces of fruit to the Slinky and repeating the steps done previously. Have them record their results and then calculate the amount of work done. Do two pieces of fruit do twice the work? Three times the work? Four times the work?

Help the students draw valid conclusions about their results. Also help them record any problems they may have encountered. For example:

- *The banana did more work than the orange.*
- *Two bananas did four times the work of one banana.*
- *The Slinky extended too far, and we could not measure the two pieces of fruit.*
- *The apple and orange weighed the same amount and did the same amount of work.*

22 | FOCUS ON MIDDLE SCHOOL PHYSICS
Teacher's Manual

Challenge Question

At the end of this *Review* section there is a challenge question. Have the students think about whether moving the Earth in this way would be possible and then help them do a rough calculation, as follows.

The mass of the Earth is 5.98×10^{24} kg.

The mass of an average human is 66 kg.

According to the US Census Bureau, the world population at the time of this writing is 7,008,096,937.

Total mass of people = number of people x mass per person
= (7,008,096,937) (66kg) = 462,534,397,842 kg ≈ 4.6×10^{11} kg.

Answer: NO — there are not enough people to move the Earth.

For up-to-date population figures go to:
http://www.census.gov/cgi-bin/ipc/popclockw

Review

(Some answers may vary.)

Define the following terms:

force *something that changes the position, shape, or speed of an object*

work *work = distance x force*

energy *gives objects the ability to do work*

In the following pairs of items, which object has the greater gravitational force? (Circle the correct answer in each pair.)

 a banana or a (bowling ball)
 a (car) or a bicycle
 the Moon or the (Earth)
 the Earth or the (Sun)

Answer the following questions:

▸ Is a book sitting on a shelf doing work? *No*

▸ Is a bowling ball crashing into the pins doing work? *Yes*

▸ How much work is done if you lift a 3 kg box 2 meters (or a 6.6 lb box 6.6 feet)? *6 kilogram-meters (43.56 ft-lbs)*

▸ How much work is done if you lift a 2 kg box 3 meters (or 4.4 lb box 9.9 feet)? *6 kilogram-meters (43.56 ft-lbs)*

List some forms of energy:

 potential energy   *chemical energy*   *electrical energy*

Challenge:

Do you think that if we could get every person on the Earth to jump all at once, we could move the Earth? Why or why not? Can you do a rough calculation to test your theory?

Chapter 3: Potential and Kinetic Energy

Overall Objectives	**24**
3.1 Potential Energy	24
3.2 A Note About Units	24
3.3 Types of Potential Energy	25
3.4 Energy Is Converted	26
3.5 Kinetic Energy	26
3.6 Kinetic Energy and Work	27
3.7 Summary	27
Experiment 3: Smashed Banana	**28**
Review	**33**

Time Required

Text reading   30 minutes
Experimental   1 hour

Materials

small to medium size toy car
stiff cardboard
wooden board
   (more than 1 meter [3 feet] long)
straight pin or tack (1-3)
small scale or balance
one banana, sliced
10 pennies
meter stick, yardstick or
   tape measure
tape

Overall Objectives

In this chapter students will be introduced to two different types of energy—potential energy and kinetic energy. Potential energy is energy that has the potential to do work, and kinetic energy is the energy of motion. The main objective of this chapter is to help students understand that energy exists in different forms and that it is converted from one form to another. Students will investigate how potential energy is converted into kinetic energy and vice versa.

3.1 Potential Energy

Potential energy is energy that has the capacity to do work but is not doing work at the moment. Have a discussion with the students about the meaning of the word potential. Make sure they understand that potential energy is already energy but that it isn't doing any work. Instead, potential energy has the capability to do work in the future. Ask the students to give some examples of potential. For example:

- *A child has the potential to become a famous scientist.*
- *A seed has the potential to become a plant.*
- *A puppy has the potential to become a dog.*

In this section of the textbook, a book on a table is used as an example of potential energy. Explain that this type of energy is formally called

- *gravitational potential energy*

Gravitational potential energy is the potential energy of an object that is elevated off the ground. It is abbreviated as GPE.

Help the students understand that the amount of GPE an object has is equal to the amount of work that was needed to lift the object to its current position. Ask the students the following questions:

- *In the Review section for Chapter 2, you calculated the work for lifting a 3 kg box 2 meters (or a 6.6 lb box 6.6 feet). How much GPE does this box have?*
  6 kilogram-meters (or 43.56 ft-lbs)

- *In the Review section for Chapter 2, you also calculated the work for lifting a 2 kg box 3 meters (or a 4.4 lb box 9.9 feet). How much GPE does this box have?*
  6 kilogram-meters (or 43.56 ft-lbs)

3.2 A Note About Units

When the students calculated work and the GPE of an elevated box, they multiplied two numbers with different units. Explain to the students that a unit describes the type of the quantity being used and that there are different kind of units. There are units for weight, such as grams and kilograms (ounces and pounds); a unit for mass—grams; units for length, such as millimeters, meters, and kilometers (inches, feet, and miles); and units for time, such as seconds, minutes, and hours.

Two different systems of units are in common use. In the United States, most people are still taught the British system of units and use this system for measuring weight, length, and volume. Units in the British system are feet, inches, miles, ounces, pounds, etc. In science, the preferred system of units is the metric system. In the metric system the units are divisible by 10, which makes calculating different quantities easier than with the British system. Metric units include meters for measuring distance, grams for mass, and liters for volume.

Have the students look at the table in this section of the student text. Point out the various equivalencies within each measurement system; i.e., how many millimeters equal a centimeter, how many meters equal a kilometer, how many inches equal a foot, how many feet equal a mile, and so on. (Note that the British units of measure in this chart are not equivalent to the metric units in the same row.) Show the students how numbers with metric units are easier to calculate than numbers in the British system. For example:

- *How many centimeters are in 5 meters?*
  (100 centimeters/meter) x (5 meters) = 500

- *How many inches are in 5 yards?*
  (12 inches/foot) x (3 feet/yard) x 5 yards = 180 inches

3.3 Types of Potential Energy

In Section 3.1 gravitational potential energy was used as an example to introduce the concept of potential energy. Other kinds of potential energy include chemical potential energy, elastic strain potential energy, and nuclear potential energy, to name a few.

Have a discussion with the students about different kinds of potential energy. For example:

- *What kind of potential energy is in cereal?*
  chemical

- *What kind of potential energy is in a battery?*
  chemical

- *What kind of potential energy is in a rubber band?*
  strain or mechanical

- *What kind of potential energy is in a spring?*
  strain or mechanical

3.4 Energy Is Converted

Have a discussion about the fact that energy gets converted. This is an important fundamental concept that the students should understand. Energy is neither created nor destroyed—only converted from one form of energy to another. It is important that the students understand that:

- *Potential energy is useful only when it gets converted to another form of energy.*

Ask the students if they can think of any useful ways to use different kinds of potential energy without first converting the energy into another form. The student textbook uses the example of making batteries into tree ornaments or jewelry. The chemical potential energy of the batteries would not be used in these examples and remains as potential energy.

3.5 Kinetic Energy

Explain that the potential energy of the book on the table (Section 3.1) gets converted into kinetic energy when the book moves from the table and begins to fall. Discuss the Greek word root for kinetic, *kinetikos,* which means "putting into motion." Explain that kinetic energy is the energy of motion.

Have the students think of some other objects that have kinetic energy. Ask them if the following things have kinetic energy:

- *a car going 45 miles/hour* (yes)
- *a tennis ball in motion* (yes)
- *a basketball sitting still on the floor* (no)
- *a toddler who is not sleeping* (yes)
- *a parent at the end of the day* (not always)

Explain to the students that the kinetic energy of an object depends on two things—the mass of the object and the speed of the object. The formula for kinetic energy (KE) is:

$$KE = 1/2\ ms^2$$

where *m* is the mass and *s* is the speed. Explain to the students that KE is proportional to both the mass of the object and its speed. This means that heavier objects will have more KE at a given speed than lighter objects, and slower objects will have less KE at a given mass than faster objects. Also notice that the KE is proportional to half of the mass and the speed squared. This means that there may be much more kinetic energy in a fast-moving toddler than his slow-moving parent!

3.6 Kinetic Energy and Work

Recall that work is simply the force of an object multiplied by the distance the object is moved. We know that an object having kinetic energy is moving, and if it hits another object, it can cause the object it hits to move. There is work done when potential energy is converted into kinetic energy and when kinetic energy is converted into other forms of energy, such as heat and sound. The work done on an object equals the change in kinetic energy of that object.

It is not important for the students to completely grasp this concept, but they should understand that when a moving object contacts another object, energy is converted and work is done.

3.7 Summary

Go over the summary statements with the students. Discuss any questions they might have.

28 | FOCUS ON MIDDLE SCHOOL PHYSICS
Teacher's Manual

In this experiment students will convert the gravitational potential energy of a small toy car into kinetic energy and do "work" on a banana.

Have the students read the entire experiment and write an objective. For example:

- *We will measure how much GPE is needed to smash a banana.*
- *We will show that a heavier toy car needs less height (less GPE) to smash a banana.*

Next, have the students write a hypothesis. For example:

- *The toy car will not be able to smash the banana no matter how high the ramp.*
- *The toy car will smash the banana when the ramp is two or three feet high.*
- *The toy car does not have enough mass to smash the banana.*
- *The toy car needs to have at least 50 pennies on it to smash the banana.*

Have the students assemble the apparatus. It helps if the board is smooth and reasonably straight. A cardboard tube, such as a wrapping paper tube, also works if cut lengthwise and opened up to make a trough. The car should have good wheels and roll smoothly and easily to reduce friction.

Experiment 3: Smashed Banana   Date: _____

Objective
_____
_____

Hypothesis
_____
_____

Materials
small to medium size toy car
stiff cardboard
wooden board (more than 1 meter [3 feet] long)
straight pin or tack
small scale or balance
1-2 bananas, sliced
10 pennies
meter stick, yardstick, or tape measure
tape

Experiment

❶ Read through all the steps of this experiment. Then write an objective and a hypothesis.

❷ Take a portion of the cardboard to make a backing for the banana slices. Using a straight pin or tack, attach a banana slice to the cardboard near the bottom.

❸ Use the wooden board to make a ramp. One end of the ramp should meet the banana slice. Your setup should look like the following illustration.

CHAPTER 3 | Potential and Kinetic Energy

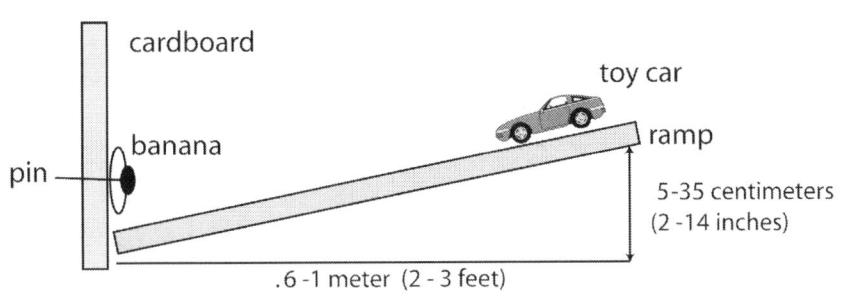

❹ Weigh the toy car with the scale or balance. Record your result.

Weight of toy car (grams or ounces) = _____

❺ Place the toy car on the ramp and elevate one end of the ramp 5 centimeters (2 inches). Allow the toy car to roll down the ramp and hit the banana. Record your results in the following chart.

❻ Elevate the ramp another 5 cm (two inches). Now the ramp should be 10 cm (4 inches) off the ground. Allow the toy car to roll down the ramp and hit the banana. Record your results in the chart below

❼ Repeat, elevating the ramp 5 centimeters (2 inches) more each time. Record your results in the following chart.

Height (centimeters or inches)	Results (write your comments)
5 centimeters (2 inches)	
10 centimeters (4 inches)	
15 centimeters (6 inches)	
20 centimeters (8 inches)	
25 centimeters (10 inches)	
30 centimeters (12 inches)	
35 centimeters (14 inches)	

When testing this experiment, we found that we needed to put several pieces of banana next to each other at the bottom of the ramp since the car often does not travel in a straight line.

Have the students weigh the toy car. Small toy cars typically weigh between 30 and 60 grams (1-2 ounces).

Have the students elevate one end of the ramp 5 cm (2 inches) and roll the toy car down the ramp. Then have them elevate the ramp in 5 cm (2 inch) increments, allowing the car to travel down the ramp and hit the banana each time the ramp is raised.

We found that an average toy car does not really smash the banana until the ramp has been elevated more than 30 cm (12 inches).

FOCUS ON MIDDLE SCHOOL PHYSICS
Teacher's Manual

Next, the students will add pennies to the car to make it heavier. They may need to tape the pennies to the car. Have them weigh the car again with the pennies on it and repeat the experiment. They should discover that the ramp will not need to be elevated quite as high in order for the toy car to smash the banana.

❽ Answer the following questions:

What happened to the speed of the car as the ramp height increased?

_____

At which ramp height did the car smash the banana?

_____

❾ Now add 10 pennies to the toy car and weigh it again. Repeat the previous steps, rolling the toy car with the pennies on it down the ramp and elevating the ramp 5 centimeters (2 inches) more each time Record your results.

Weight of toy car plus 10 pennies (grams or ounces) = _____

Height (centimeters or inches)	Results (write your comments)
5 centimeters (2 inches)	
10 centimeters (4 inches)	
15 centimeters (6 inches)	
20 centimeters (8 inches)	
25 centimeters (10 inches)	
30 centimeters (12 inches)	
35 centimeters (14 inches)	

❿ Answer the following questions:

At which ramp height did the car smash the banana? _____

Was the banana smashed at the same height by the light car and the heavy car? _____

If "no," which car needed to be at a greater height to smash the banana? _____

Results

Calculate the GPE for the height of the ramp at which the toy car—with and without the 10 pennies—smashed the banana. Use the equation:

gravitational potential energy (GPE) = weight × height

Record your answers below.

GPE for car without pennies _____

GPE for car with pennies _____

Is the GPE the same (or close to the same) for both cars? _____

Have the students calculate the GPE for the toy car with and without the pennies at the corresponding heights at which the banana was smashed.

The GPEs should be roughly equal. Basically, we expect that it takes a given amount of KE (kinetic energy) to smash the banana, and it doesn't matter whether this comes in the form of a heavy, slow car or a light, fast car. The GPE the students calculate is the energy needed to smash the banana.

Have the students write valid conclusions based on the data they have collected.

Conclusions

CHAPTER 3 | Potential and Kinetic Energy

Review

(Some answers may vary)

Define the following terms:

potential energy *energy that has the potential to do work*

gravitational potential energy *the potential energy of an object that is elevated off the ground*

chemical potential energy *potential energy stored in chemicals*

kinetic energy *the energy of moving objects*

Fill in the blanks for the following measurements:

1 foot = *12* inches
1 yard = *3* feet
1 mile = *1760* yards

1 meter = *100* centimeters
1 centimeter = *100* millimeters

1 gram = *0.001* kilogram
1 pound = *16* ounces

Chapter 4: Motion

Overall Objectives	35
4.1 Motion	35
4.2 Inertia	36
4.3 Mass	36
4.4 Friction	37
4.5 Momentum	37
4.6 Summary	38
Experiment 4: Moving Marbles	39
Review	44

Time Required

Text reading   30 minutes
Experimental   1 hour

Materials

several glass marbles of different sizes
steel marbles of different sizes
cardboard tube, .7-1 meter [2.5-3 ft] long
scissors
black marking pen
ruler
letter scale or other small scale or balance

Overall Objectives

In this chapter students will learn about some properties of motion: inertia, friction, and momentum. It is important for the students to understand that an object will remain in a steady, straight line of motion until a force acts on it (review force in Chapter 2, Section 2.2 of the student textbook).

4.1 Motion

Have the students observe or think about objects that move, and ask them why they think objects move. Revisit the Chapter 1 experiment, *It's the Law!* Ask what they discovered about objects that move by doing that experiment and have them reread their conclusions. Explain that what they observed in the experiment was *Newton's First Law of Motion,* which states:

- *An object in motion will stay in motion unless acted on by an outside force, and an object at rest will stay at rest unless acted on by an outside force.*

Have a discussion with the students about the significance of this statement. Explain that for 2000 years people thought that an object that was moving had to have a force pushing it. Aristotle thought that this was how objects moved, but he was wrong. Because Aristotle did not feel the movement of the Earth but saw the Sun and Moon move in the sky, he thought that the Earth was the center and that the Sun and Moon moved around the Earth. Discuss the Greek word roots for *geocentric cosmos.*

Next, have the students look at the drawing of the solar system in the textbook. Show them where the Earth and Sun are with respect to each other. Explain that we know today that we live in a heliocentric cosmos in which the Earth rotates around the Sun. Discuss the Greek word roots for *heliocentric cosmos.*

Explain to the students that there were early scientists who challenged the idea of a geocentric cosmos, but it took 2000 years for people to finally believe that the Earth was not the center of the universe. Ask the students why they think this might have happened. Discuss the fact that there are many different factors that go into scientific discovery and that, because scientists are human, there are personal biases and pressures that occur. It is sometimes difficult to replace a prevailing scientific theory with a new theory because of these biases and pressures. This still happens today, and new theories are not readily accepted if they are a major challenge to the dominant paradigm.

4.2 Inertia

Lead a discussion about the concept of *inertia,* which is the tendency of things to resist a change in motion. Ask the students what they think this means.

Ask them if the following objects are easy or hard to move:

- *a toy boat floating on water*
- *a rowboat on water*
- *a 20-foot motor boat on water*
- *an ocean liner on water*

Explain that in physics there are two aspects to consider with regard to inertia:

- *mass (see Section 4.3)*
- *momentum (see Section 4.5)*

4.3 Mass

It is important for the students to understand that *mass and weight are different*. Weight is a force. Mass is not. However, by weighing an object you can tell how much mass it has. The more mass an object has, the more it will weigh on Earth because more force will be exerted on it than on an object with less mass. Explain to the students that without gravity, objects do not weigh anything. In space, where there is no gravity, a boulder would float in the same way a feather would. However, in space, the boulder and the feather would still have different masses. The boulder still has more mass than the feather and, as a result, would still be harder to accelerate (speed up) than a feather—even in space.

Objects with a large mass *accelerate* slowly when pushed by a force. If you push on a small object that has little mass, such as a toy boat, it accelerates quickly. On the other hand, if you push a larger object, like a good-sized sporting boat, it will eventually begin to move, but you will have to push for a longer time before it accelerates. Explain to the students that, in the absence of friction, they could even move a large-size ocean liner like the Queen Mary by pushing on it, but it would take time to start it moving. Here, we are considering inertia with reference to an object's mass.

4.4 Friction

Review Newton's First Law of Motion: *An object in motion will stay in motion unless acted on by an outside force, and an object at rest will stay at rest unless acted on by an outside force.*

Although inertia keeps things moving, objects on Earth will eventually stop. Ask the students:

- *What is the only reason why any object would stop moving?*
  (because a force acts on the object)

Then ask what force is acting on the following objects to make them stop moving, and ask if this force can be "seen."

- *a rolling marble*
- *a hockey puck on ice*
- *a car out of gas*

Talk about the fact that, even though the students cannot visibly see why these objects stop moving, a force is still acting on them. Otherwise the objects would never slow down or stop.

This force is *friction*. Friction occurs when two objects rub against each other. Have a discussion with the students about friction being the force that works in the direction opposite the direction of motion. Friction is what slows objects down and eventually causes them to stop. Explain that, in the absence of friction, an object would keep moving forever and never stop.

In the examples above: for the marble, there is rolling friction between the floor and the marble; for the hockey puck, there is slight friction between the puck and the ice; for the car, most of the friction is within the engine, wheels, and axles.

4.5 Momentum

The second aspect of inertia is the fact that an object with a lot of momentum is hard to stop. *Momentum* is inertia in motion; that is, mass that is moving. Momentum makes objects hard to stop. Recall that *inertia* is the tendency of an object to resist a change in motion. Objects that are stationary want to remain stationary, and objects that are moving want to remain in motion.

The mathematical equation for momentum is:

$$\text{momentum} = \text{mass} \times \text{speed}$$

Explain to the students that an object that has a large mass will have a large momentum. Also, an object moving at a fast rate of speed will have a large momentum.

Ask the students if they could stop these objects with their bare hands without feeling pain:

- *a baseball tossed in the air*
  (yes)
- *a basketball thrown to them*
  (yes)

Now ask them if they could stop the following objects with their bare hands without feeling pain:

- *a baseball hit with a bat*
  (No)
- *a basketball shot out of a cannon*
  (No)

Ask them *why*.

- *Have the masses of these two objects changed?*
  (No)
- *What is different?*
  (The objects are traveling at faster speeds in the second examples.)

4.6 Summary

Go over the summary statements with the students. Discuss any questions they may have.

Experiment 4: Moving Marbles   Date:_____

Objective _____

_____

Hypothesis _____

_____

Materials

- several glass marbles of different sizes
- steel marbles of different sizes
- cardboard tube, .7-1 meter (2 1/2-3 ft) long
- scissors
- black marking pen
- ruler
- letter scale or other small scale or balance

Experiment

1. Using the scale, weigh each of the marbles, both glass and steel. Label the marbles with numbers or letters, or note their colors, so that you can keep track of how much each marble weighs. Record your results in **Part A** of the *Results* section.

2. Take the cardboard tube and cut it in half lengthwise to make a trough. Measure the length of the tube, and mark the halfway point with the black marking pen.

3. Beginning at the halfway mark, measure .3 meter (1 foot) in both directions, and put a mark at each of these measurements. This will give you one mark on each side of the halfway mark.

4. The cardboard tube should now have three marks: one at the halfway point, and one on either side of the halfway mark, .3 meter (1 foot) away from it. The tube will be used as a track for the marbles.

Have the students read the entire experiment and write an objective. Some possible objectives are:

- *We will examine the movement of different marbles.*
- *We will investigate the momentum of different marbles.*
- *We will see what happens when one marble hits another.*
- *We will see if we can move a heavy marble with a light one.*
- *We will see if we can move a light marble with a heavy one.*

Have them write a hypothesis. Some examples are:

- *The small glass marble will not be able to move the steel marble.*
- *The small glass marble will be able to move the steel marble.*
- *The small glass marble will stop when it hits the steel marble.*
- *The small glass marble will not stop when it hits the steel marble.*

In Step 1, students will weigh the marbles. Remind them that weight and mass are different and that they are not going to find the actual mass of the objects. However, they will be able to tell which objects have more mass—that is, those that weigh more.

Have the students mark the marbles or write down the colors of the marbles so that they will know which marble corresponds to each weight. Have them record the description and weight of each marble in Part A of the *Results* section.

Have the students cut and mark the cardboard tube as described in Steps 2-4.

40 | FOCUS ON MIDDLE SCHOOL PHYSICS
Teacher's Manual

In Step 5, the students will roll the marbles down the tube one at a time and observe how each marble rolls. Have them describe the movement of the marbles in Part B of the *Results* section. Some examples are:

- *The glass marbles move easily down the tube and off the end.*
- *The small steel marbles move easily down the tube.*
- *The large steel marble takes more effort to get it to move down the tube.*

They may notice that it takes slightly less effort to push the glass marbles than the heavy steel marbles. This is because the larger steel marbles have more inertia than the smaller marbles.

In Steps 6 and 7, they will place a glass marble in the center of the tube. When they roll another glass marble of the same size down the tube, they should watch closely to see what happens. Have them record their observations in Part C. Some example observations are:

- *The rolling glass marble hit the other marble and stopped.*
- *The marble that was stopped started moving when it was hit by the rolling glass marble.*

  (Answers will vary.)

Ask the students to observe the following:

- *When you roll the glass marble slowly, how far does the marble it hits move?*
- *When you roll the glass marble fast, how far does the marble it hits move?*

❺ Take the marbles and, one by one, roll them down the tube. Notice how each one rolls (Does it roll straight? Is it easy to push off with your thumb? Does it pass the marks you drew?) In **Part B** of the *Results* section, describe how each marble rolls.

❻ Now place a glass marble on the center mark of the tube.

❼ Roll a glass marble of the same size toward the marble in the center. Watch the two marbles as they collide. Record your results in the *Results* section, **Part C**.

❽ Repeat Steps 6 and 7 with different size marbles. For example, try rolling a heavy marble toward a light marble and a light marble toward a heavy marble. Record your results in **Part D**.

Results

Part A

Marble	Weight

Part B

_____
_____
_____
_____
_____
_____
_____
_____

What they are observing here is the *conservation of momentum*—the total *linear momentum* (mass x speed) stays the same. However, because the rolling marble also has *angular momentum* (converted from linear momentum due to rotation) in addition to *linear momentum,* and also experiences *friction*, the stationary marble does not pick up quite all of the total linear momentum of the moving one. In the absence of angular momentum and friction, the total linear momentum of the moving marble would be transferred to the stationary marble. Despite this conversion of some of the linear momentum, the students should still be able to observe qualitatively that, when hit, the stationary marble will move faster when the rolling marble is traveling faster.

A better way to illustrate this would be to use an air hockey table. If one is available, try the same experiment with two hockey pucks. The advantages to using an air table are: first, the pucks have little friction, so momentum is better conserved; and second, the pucks are not rolling, so there is no angular momentum to complicate things.

Next, have the students use different size marbles. Have them roll a light marble toward a heavy marble and a heavy marble toward a light marble. Help them carefully observe what happens. They should observe that:

- *When the heavy marble impacts the light marble, the light marble will accelerate quite a bit.*

- *When the light marble impacts the heavy marble, the heavy marble will accelerate only a little bit.*

42 | FOCUS ON MIDDLE SCHOOL PHYSICS
Teacher's Manual

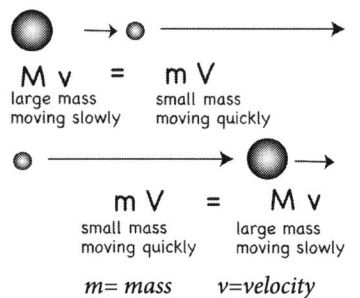

m = mass v = velocity

Have the students repeat this several times. Ask them what they think of these results.

Discuss the *conservation of momentum* which states that the total momentum stays the same.

Remind the students that momentum is mass x speed (velocity). A large mass traveling at a certain speed will cause a smaller mass to accelerate to a faster speed, and a small mass traveling at a fast speed will accelerate a large mass to a slower speed. Because momentum is conserved, the total momentum will stay the same in each case.

Although there is angular momentum in a rolling ball and friction is also present, the conservation of momentum should again be observable qualitatively in this part of the experiment. Have the students draw conclusions based on the data they have collected.

Part C

_____
_____
_____
_____
_____
_____
_____
_____
_____

Part D

_____
_____
_____
_____
_____
_____
_____
_____
_____
_____
_____

Conclusions

Review

(Some answers may vary)

Define the following terms:

inertia   *the tendency of objects to resist changes in motion*

mass   *the property that gives objects inertia*

momentum   *the property that makes objects hard to stop*

friction   *the force experienced by two objects rubbing against each other*

▶ Write the equation for momentum:

momentum = mass x speed

▶ Which has more mass: (a bowling ball) or a green pea? (circle one)

▶ Which has more momentum: a rolling bowling ball or (a bowling ball shot from a cannon)? (circle one)

Chapter 5: Energy of Atoms and Molecules

Overall Objectives	46
5.1 Chemical Energy	46
5.2 Stored Chemical Energy	46
5.3 Stored Chemical Energy in Food	47
5.4 Stored Chemical Energy in Batteries	47
5.5 Nuclear Energy	48
5.6 Summary	48
Experiment 5: Power Pennies	49
Review	54

Time Required

Text reading   30 minutes
Experimental   1 hour

Materials

10-20 copper pennies
aluminum foil
paper towels
salt water (30-45 ml [2-3 Tbsp] salt per 240 ml [1 cup])
voltmeter
2 plastic-coated copper wires, each 4"-6" long
duct tape or other strong tape

scissors
wire cutters
steel wool

Optional Materials

vinegar, 240 ml (1 cup)

Overall Objectives

In this chapter students will learn about the energy of atoms and molecules; that is, chemical potential energy. Chemical potential energy is found in things such as fuels, foods, and batteries. Students will also be introduced to nuclear energy. It is important to help the students understand that all forms of chemical potential energy are useful only when they are converted to other forms of energy, such as light, heat, or electricity.

5.1 Chemical Energy

Discuss chemical energy with the students and have them think about some different chemical reactions they may be familiar with. After reading the textbook, have them look at the illustration in this section and discuss what is happening with the jug, vinegar, and baking soda. Explain that as the chemical reaction occurs in the jug causing the cork to pop off, chemical energy is being converted into mechanical and kinetic energy. Ask them if they can think of other chemical reactions that get converted into other forms of energy. For example:

- *gasoline in a lawn mower* (chemical to mechanical)
- *chemical heating pack* (chemical to heat)
- *chemical cooling pack* (chemical to heat loss)
- *chemical light stick* (chemical to light)
- *matches* (chemical to heat)

5.2 Stored Chemical Energy

Discuss the fact that chemical energy starts as *chemical potential energy* before it is converted into other forms of energy. Have the students look closely at the illustration in the textbook that shows a steam engine fueled by wood. Explain that, in this illustration, the *stored chemical energy* in the wood and oxygen gets converted into *heat energy* when the wood burns. The heat energy is used to heat water which expands as steam. The work done by the expanding steam is converted into *mechanical energy* as the piston moves up and down and the train moves forward. In this illustration the energy stored as chemical potential energy is used to convert several different forms of energy from one to another. Have the students look up how cars use gasoline and discuss any similarities or differences.

Similarities:

- *fuel (gasoline) is burned to produce heat and expanding gases*
- *pistons are used*
- *the chemical energy (in the gasoline and oxygen) is converted into mechanical energy*

Differences:

- *water is not heated in a gasoline engine*
- *the air in a gasoline engine is compressed*
- *the gasoline is burned inside the cylinders, not outside*

5.3 Stored Chemical Energy in Food

Have a discussion about chemical energy that is stored in food. Have the students look at Chapter 8 in the *Focus On Middle School Chemistry Student Textbook* and review the chemical energy found in certain foods. Talk about which foods have carbohydrates. For example:

- *potatoes*
- *bread*
- *sugar*
- *spaghetti*

Explain to the students that the body gets its energy from stored chemical energy in foods. The chemical reactions that occur inside the body are complicated, but they amount to essentially the same thing as burning sugar. Many different kinds of reactions take place but, overall, the foods we eat get burned and turned into other forms of energy, such as heat and mechanical energy (mainly heat). Our bodies require a continual supply of energy foods because, unlike plants, we cannot produce our own food. Ask the students:

- *What is the one source of energy on which all plants and animals depend?"* (the Sun)

5.4 Stored Chemical Energy in Batteries

Another form of stored chemical energy is found in batteries. Explain that a battery is specifically designed to convert chemical energy into electrical energy. Ask the students to list familiar items that use batteries. Some examples:

- *cell phone*
- *flashlight*
- *many different battery-operated toys*
- *automobiles*

The first battery was invented by Alessandro Volta. He constructed a voltaic cell which used alternating layers of metals and salt water to generate electrical energy. A different type of battery called a dry cell is used in cell phones and flashlights (see Chapter 6).

5.5 Nuclear Energy

Have a discussion about *nuclear energy,* which is a form of stored energy in atoms. Nuclear energy is released when the nuclei of atoms split into smaller pieces or when smaller nuclei combine into bigger ones. Explain that nuclear reactions differ significantly from chemical reactions. In nuclear reactions the atoms themselves change their identities—for instance, from carbon (C) to nitrogen (N). In chemical reactions the atoms keep their identities and only change which other atoms they are bonded to.

Discuss the ways in which an atom can change. Explain that, by changing the number of protons or neutrons in an atom's nucleus, the element changes. Discuss the illustration in the textbook that shows how a nitrogen atom gets converted into a carbon atom by losing a proton. The carbon atom is called carbon 14 (14 = 6 protons + 8 neutrons). Have the students look at the periodic table found in the *Focus On Middle School Chemistry Student Textbook,* and ask them how many protons and neutrons a carbon atom has. Point out that a carbon 14 atom still has 6 protons, but has 8 neutrons, two more than normal carbon. This makes carbon 14 an *isotope* of carbon. An isotope has the same number of protons as the normal element but a different number of neutrons.

Nuclear reactions release much more energy than chemical reactions, and they are used to power nuclear reactors. Use the illustration in the textbook to explain the design of a nuclear reactor. Point out that the nuclear energy produces heat energy which is then used to heat water. The steam from the water turns a turbine that generates electrical energy.

5.6 Summary

Go over the summary statements with the students. Discuss any questions they might have.

Experiment 5: Power Pennies   Date: _____

Objective _____

_____

Hypothesis _____

_____

Materials

- 10-20 copper pennies
- aluminum foil
- paper towels
- salt water: 30-45 ml (2-3 Tbsp.) salt per 240 ml (1 cup) water
- voltmeter
- 2 plastic-coated copper wires, each 10-15 cm (4"-6") long
- duct tape (or other strong tape)
- scissors
- wire cutters
- steel wool

Experiment

❶ Cut out several penny-sized circles from the aluminum foil and a paper towel.

❷ Soak the paper towel circles in the salt water.

❸ Strip the plastic coating off both ends of one of the pieces of wire, using wire cutters to carefully cut through the plastic without cutting the metal wire. Take one end of the wire and tape the exposed metal to a penny.

❹ Strip the plastic off the ends of the other piece of wire. Tape the exposed metal on one end of the wire to a piece of aluminum foil.

CHAPTER 5 | 49
Energy of Atoms and Molecules

Have the students read the entire experiment before writing the objective and hypothesis. Some examples are:

Objective

- *To discover how a simple voltaic cell operates.*
- *To construct a voltaic cell and measure voltages.*
- *To see if pennies, aluminum, and salt water can really make electricity.*

Hypothesis

- *Pennies, aluminum foil, and salt water will not generate electricity.*
- *Pennies, aluminum foil, and salt water will generate electricity.*
- *More layers in the voltaic cell will generate more electricity.*

Assemble all of the materials before starting. It helps to scrub the pennies with steel wool. Help the students cut out small, penny-sized circles of aluminum foil and paper towel. It is important that the cutouts be very close to the size of a penny. Have them soak the paper circles in the salt water.

Help the students carefully strip the plastic off both ends of each piece of wire. It can be difficult to strip just the plastic and not cut the wire itself. The best way is to use a pair of wire cutters. Gently squeeze the wire cutters and pull the plastic off the end of the wire.

Have the students tape the exposed end of one wire to a penny and the exposed end of the other wire to one circle of aluminum foil. These will remain as the ends of the battery. Additional layers will be added in between these two ends.

Now have the students carefully insert one of the salt water soaked paper circles in between the copper penny and aluminum foil. It helps to set the aluminum foil circle wire side down on a firm surface, add the salty paper towel circle, and then place the penny, wire side up, on top, holding it down with the fingers.

Have the students hook the leads (wires from the battery) to a voltmeter and read the voltage. An inexpensive voltmeter can be purchased at any store that supplies electrical equipment. Carefully read the instructions for the voltmeter. Make sure that the voltmeter is set to "voltage" and that the voltage scale is low enough to detect small voltages. A typical penny-cell produces about 0.5v.

❺ Begin stacking the pieces by placing the circle of aluminum foil with the wire attached wire side down on a firm surface. Put one of the wet paper towel circles on top of the aluminum foil. On top of the paper, place the penny that has the wire taped to it. It should look like this:

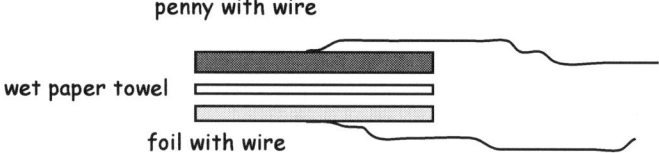

❻ Take the wires and connect them to the leads (wires) of the voltmeter. Switch the voltmeter to "voltage" and record the number it shows. This is the amount of voltage the single layer battery produces.

❼ Add another "cell" to the battery and record the voltage. (A cell is a penny layer, a paper layer, and a foil layer.) The battery now has two cells. It should look like the following:

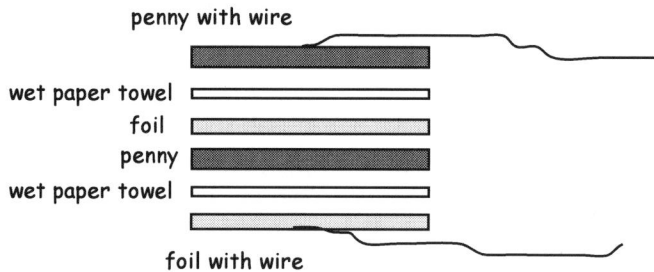

❽ Continue adding cells made of foil, wet paper towel, and pennies, and record the voltage when each new cell is added.

Results

Number of Cells	Voltage
1	
2	
3	
4	
5	

Have the students record the voltage they read on the voltmeter. Next have them add additional "cells" to the battery. A cell consists of an aluminum foil layer, a soaked paper layer, and a penny layer.

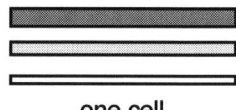

one cell

The students should add as many additional cells as they can and record the voltage each time a new cell is added.

Have the students plot their data. The voltage should be on the x-axis (horizontal) and the number of cells on the y-axis (vertical).

See the *Focus On Middle School Chemistry Student Textbook*, Section 5.4, and the *Focus On Middle School Chemistry Teacher's Manual*, Chapter 5, for information about plotting data.

Plot your data. Make a graph with "Voltage" on the x-axis and "Number of Cells" on the y-axis.

Discuss your data.

Conclusions

Have the students discuss their data. They should observe that, overall, as the number of cells increases, the voltage increases. However, if there are places where the paper towel or aluminum foil is enough larger than the penny that it hangs down to touch a lower layer, there will be a short circuit—that is, the electricity will travel through the shortest distance, missing parts of the cells. If this happens, the voltage will fluctuate—appear to increase and then decrease.

paper towels touch

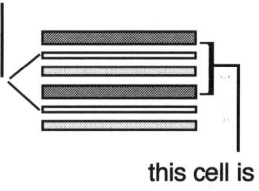

this cell is "short circuited"

Have the students discuss possible sources of error.

Have them draw conclusions based on the data they have collected.

Additional Experiment:

It is also possible to use vinegar instead of salt water. If time permits, have the students repeat the experiment using vinegar and compare the results to those of the salt water based battery.

Review

Answer the following questions:

▸ What is chemical energy?   *the energy released from chemical reactions*

▸ Name two foods that have "food energy" (carbohydrates).

   potatoes

   pasta

▸ What are two examples of energy we use for fuel?

   gasoline (wood, coal, sunlight, etc.)

   nuclear reactions

▸ Who made the first battery?

   Alessandro Volta

Draw a diagram of a voltaic battery.

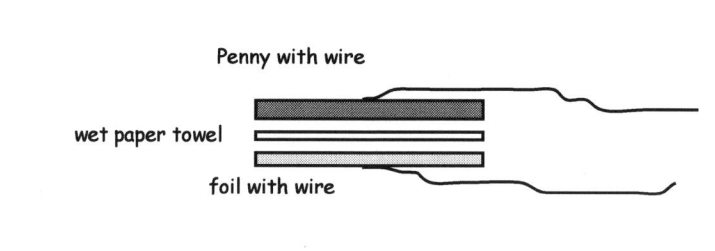

Chapter 6: Electrical Energy and Charge

Overall Objectives	**56**
6.1 Electrical Energy	**56**
6.2 Electric Charge	**56**
6.3 Charging Objects	**57**
6.4 Electrical Force	**57**
6.5 Summary	**57**
Experiment 6: Charge It!	**58**
Review	**62**

Time Required

Text reading 30 minutes
Experimental 1 hour

Materials

small glass jar with lid
aluminum foil
paper clip
duct tape (or other strong tape)
plastic or rubber rod (or balloon)
silk fabric (or hair—see Step 7)
scissors
ruler

Optional Materials

several dry cell batteries of different sizes and shapes for observation

Overall Objectives

In this chapter students will learn about electrical energy. They will review the structure of an atom and examine in more detail the nature of electric charge.

6.1 Electrical Energy

In Chapter 5 students were introduced to electrical energy that is produced with batteries. In the experiment for Chapter 5, the students built a voltaic battery and discovered that as they added more cells, more voltage was produced.

Have a discussion with the students about other types of batteries. The type of battery used in most small electronic equipment is called a dry cell battery. Have the students look at some dry cell batteries. DO NOT allow the students to break open a dry cell battery. These batteries contain caustic materials that can burn skin.

Have the students look at different sizes and shapes of dry cell batteries and compare the voltages. Ask them why they sometimes need to put more than one battery in a toy or CD player. Ask them what they notice about how the batteries are positioned in a typical electronic toy.

Ask the students if there are other objects that require electrical energy but that do not use batteries. For example:

- *microwave oven*
- *washing machine*
- *hair dryer*
- *power saw*

Ask the students if they are aware of other forms of electrical energy (static electricity, lightning, etc.).

6.2 Electric Charge

Have the students review the parts of an atom. It would be helpful to use Chapters 1 and 2 in the *Focus On Middle School Chemistry Student Textbook* for the review. Explain that the parts of an atom that "carry" electric charges are the electrons and protons—neutrons have no charge. Scientists have arbitrarily assigned the positive sign (+) to represent the charge of protons and the negative sign (-) to electrons. That is, by convention protons are positive and electrons are negative. The important aspect is that they are oppositely charged.

Explain to the students that:

- *opposite charges attract*
- *like charges repel*

Have them look closely at the diagram of the atom in the textbook. Point out that a helium atom has two electrons and two protons. Explain that the two electrons in a helium atom will try to be as far away from each other as possible. It is the positive charges of the protons that keep the negatively charged electrons from flying away from the atom.

6.3 Charging Objects

Have the students list some times when they have observed objects being charged. For example:

- *pulling clothes out of the dryer*
- *rubbing a balloon in hair and sticking it to a wall*
- *dragging their stocking feet on the carpet*

Explain that in all of these cases objects are being charged by electrons being moved from one object to another. The fact that the objects are charged becomes visible when two objects are observed "sticking" to each other (such as socks and bedsheets coming from a dryer and sticking together), or when objects repel each other (such as individual hairs that are charged with static electricity separating themselves from each other and sticking outward from the head).

6.4 Electrical Force

Discuss electrical force with the students. Explain that electrical force is like other forces because electrical force causes a change in the shape or speed of something, such as the movement of charged particles like electrons.

Ask the students to list some ways electrical forces change the position of an object. Some examples are:

- *A charged balloon will pull objects, such as oppositely charged hair, toward it.*
- *A charged bedsheet will drag a sock out of the dryer.*
- *Some types of plastic peanuts become irritatingly charged and difficult to remove from hands and other surfaces.*

Ask the students if they think they could make a simple instrument to detect electrical charge. In the experimental section of the workbook, the students will construct a simple electroscope and will detect electrical charge.

6.5 Summary

Discuss the summary statements with the students.

58 | FOCUS ON MIDDLE SCHOOL PHYSICS
Teacher's Manual

Have the students read the entire experiment and then have them write an objective and a hypothesis.

Some examples are:

Objective

- We will build an instrument to detect electric charge.
- We will test for electric charge with an electroscope.

Have the students guess what might happen with the electroscope. Give them the hint that the two pieces of aluminum foil will carry the same charge. Then ask what the aluminum foil will do.

Hypothesis

- The aluminum foil pieces will separate in the electroscope.
- The aluminum foil pieces will stick to each other in the electroscope.

Have the students assemble the parts to make the electroscope. Provide assistance as they puncture a small hole in the top of the jar lid.

Have them attach the two aluminum foil pieces to the paper clip hook and then insert the hook and aluminum foil into the jar by placing the lid on the jar.

Experiment 6: Charge It!   Date: _____

Objective _____
_____

Hypothesis _____
_____

Materials

small glass jar with lid
aluminum foil
paper clip
duct tape (or other strong tape)
plastic or rubber rod (or balloon)
silk fabric
scissors
ruler

Experiment

Building an *electroscope* (an instrument that detects electric charge)

❶ Cut two thin strips of aluminum foil of equal length (about 2.5 cm [1 inch] long).

❷ Poke a small hole in the center of the lid of the glass jar.

❸ Open one end of the paper clip to make a small hook.

❹ Place the straightened out piece of the paper clip through the small hole in the jar lid, bend, and secure the paper clip to the lid with strong tape, leaving the end of the paper clip exposed.

❺ Hang the two strips of aluminum foil from the hook that is on the underside of the jar lid. Place the lid on the jar with the aluminum foil hanging from the hook inside the jar.

❻ You now have an electroscope.

❼ Take the plastic or rubber rod and rub it with the silk fabric, or take the balloon and rub it in your hair or on the cat.

❽ Gently touch the plastic or rubber rod or the balloon to the end of the paper clip that is sticking out of the glass lid.

❾ Observe the two pieces of aluminum foil and record your results.

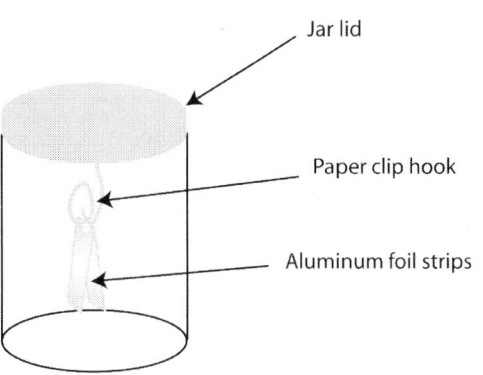

Jar lid
Paper clip hook
Aluminum foil strips

Once the electroscope is assembled, have the students rub a plastic rod (a plastic comb will work) against a piece of silk cloth or in their hair. They should be able to charge the plastic rod or comb in this way.

Have them gently touch the rod to the tip of the paper clip that is sticking out from the jar.

Have them observe what is happening to the two aluminum foil pieces. They should see them separate. Have them record what they see. Have them think about the following:

- *What would happen if you touched your fingers to the end of the paper clip?*

- *How long will the charge last?*

- *Are there things that would make the aluminum foil separate farther? More rubbing? Different plastic rod? A glass rod? A metal rod?*

How the electroscope works:

Like charges repel, so with a charged object like a balloon (or the parts of an electroscope) the charges always spread out as far apart as they can. When you touch a charged object like the plastic rod to the paper clip in the electroscope, the electrons spread from the rod through the paper clip and the aluminum strips. Now both of the foil strips have negative charges, so they repel.

After awhile the charge leaks away, and the strips come back together. The bigger the charge, the more the strips repel; therefore, you can tell how strongly charged the rod was to start with.

charged rod

before touching

charged rod

electric charge flows from the rod to the paper clip and the aluminum foil

less charge on the rod

aluminum foil strips have "like" charge, so they repel

Results

Conclusions

Help the students write valid conclusions based on the data they have collected.

Review

(Some answers may vary.)

Define the following terms:

dry cell — *a battery that use pastes instead of liquids to convert chemical energy to electrical energy*

electric charge — *the charge assigned to protons and electrons*

electrical force — *force that causes the movement of charged particles*

friction — *the force that makes objects slow down and stop*

Circle the correct word to complete the statement:

▸ Like charges (**repel** / attract) each other.

▸ Unlike charges (repel / **attract**) each other.

List the parts of an atom and whether or not they are charged.

proton — positively charged

neutron — no charge

electron — negatively charged

Chapter 7: Moving Electric Charges and Heat

Overall Objectives	64
7.1 Moving Electric Charges	64
7.2 Resistance	65
7.3 Heat	65
7.4 Summary	66
Experiment 7: Let It Flow	67
Review	71

Time Required

Text reading   30 minutes
Experimental   1 hour

Materials

1.2 meters (4 ft) insulated electrical wire
6v or larger (up to 12v) battery
insulating materials
    (e.g., styrofoam, plastic, cloth)
small light bulb (Radio Shack flashlight
    lamp #272-1163 with a rated voltage of 6v
    —or comparable bulb)
electrical tape
several small resistors
scissors
wire cutters

Optional: 2 alligator clips

Overall Objectives

In this chapter students will learn about electric current, resistance, and heat. It is important for the students to understand that charges move, or flow, through wires as a result of electric pressure. This pressure is the result of a difference in electric potential (or voltage) in an electric circuit.

7.1 Moving Electric Charges

Explain to the students that in the last chapter they observed the effects of static electric charges in the electroscope. Static charges can be transferred from object to object, but once they are in place, static charges do not flow.

Ask the students the following questions:

- *How long does it take for a light bulb to light up once the switch is turned on?*
  (lights up immediately)

- *How long does it take for the TV to turn on when you plug it in?*
  (turns on immediately)

- *What if you take an extension cord and make the cord very long. How long will it take to turn on a power drill?*
  (turns on immediately—even with a long extension cord)

- *Does the length of the cord matter?*
  (No)

Have a discussion with the students about electric current. When charges move, there is an electric current, and this current behaves much like a hose filled with water. The pressure that the water hose experiences when the faucet is turned on causes the water to be pushed out of the hose. In fact, if the hose is completely filled with water when the faucet is turned on, water will immediately flow out the other end of the hose. This is similar to how electric current works.

Explain that electric pressure is called voltage. The higher the voltage, the more electric pressure, and the more electrons can be moved.

Explain to the students that wires are made of metals, and metals conduct electricity; that is, metals contain many electrons that can move. When electric pressure is applied to a wire, the electrons inside the wire move. This is why the length of the cord does not matter. Electric pressure at one end pushes electrons out the other end immediately. So the light bulb, TV, and power drill all turn on immediately when the cord is plugged into the wall.

Have the students look at this section of the textbook and examine the diagram showing several atoms of conducting material. Point out that the electrons move, but the protons and neutrons do not.

7.2 Resistance

Materials that allow electrons to flow easily are called conductors. Metals are good conductors because they have lots of electrons that are free to move from atom to atom. Insulators are materials that don't allow electrons to move easily from atom to atom. Insulators are resistant to electric flow.

Electrons can move from atom to atom only if there is a place in the receiving atom to accommodate the incoming electron. Insulators do not have space in their atomic shells to accommodate extra electrons, so their electrons do not move.

Explain to the students that in most electrical circuits there are small components called resistors. Resistors are used to slow down the flow of electrons, controlling how much electric current flows.

Typical resistors look like this:

The stripes show the resistance value of the resistor.

7.3 Heat

When lots of electric current flows through a wire, the wire often feels warm or hot to the touch.

Ask the students to describe some things that are heated electrically. For example:

- *burners on an electric stove top*
- *filament in an incandescent light bulb*
- *toaster*
- *electric blanket*

Ask the students where they think the heat comes from in these cases.

Explain that when electrons flow through a metal, the electrons collide with atoms and impart some kinetic energy to the atoms. The atoms shake and vibrate, and we experience this extra energy as heat. The greater the electric current that flows through the metal, the more the metal is heated. Some materials get so hot they glow, like a red-hot burner on a stove top or the white-hot filament in an incandescent light bulb.

Have the students list some objects that do not get hot. For example:

- *styrofoam*
- *certain plastics*
- *wood*
- *cloth*

Explain to the students that these are all insulators—they do not conduct electricity, and no electric current flows through them.

7.4 Summary

Review the summary statements with the students.

Experiment 7: Let It Flow

Date: _____

Objective
_____
_____

Hypothesis
_____
_____

Materials

- 1.2 meters (4 ft) insulated electrical wire
- 6v or larger battery
- insulating materials (such as styrofoam, plastic, cloth, etc.)
- small light bulb
- electrical tape
- several small resistors
- scissors
- wire cutters

Experiment

1. Cut the wire into two pieces, each about .3 meter (1 foot) long. Carefully shave the ends of the plastic insulation from the ends of the wires to expose the metal. Leave about 6-12 mm (1/4 to 1/2 inch) of exposed metal on each end.

2. Tape one end of one wire to the positive (+) terminal of the battery. Tape one end of the other wire to the negative (-) terminal of the battery. Alligator clips can be used instead of electrical tape to fasten the wires to the battery terminals.

3. Tape the other ends of the two wires to the light bulb. One wire should be taped to the bottom of the bulb, and the other one should be taped to the metal side of the bulb.

   Record your results on the chart in the *Results* section.

Have the students read the entire experiment and then write an objective and a hypothesis. Some examples are:

Objective

- *We will find out if materials such as foam and plastic conduct electricity.*
- *We will observe electric current and find out what happens if an insulating material interrupts the flow.*

Hypothesis

- *The insulators will not affect the electric flow.*
- *The insulators will keep the light bulb from lighting.*

Have the students cut two pieces of wire and carefully strip the plastic off both ends of each, exposing the metal. Then they will either tape or wrap one end of each piece of wire around a terminal of the battery. Alligator clips can provide a more secure way of fastening the wires to the battery. If these are used, help the students fasten the wires to the alligator clips.

Next, have them tape the other ends of the wires to a small light bulb. One wire should be taped to the bottom of the bulb, and the other should be taped to the metal side. It should look like this illustration:

Have the students follow all the steps of the experiment, record the results of each step, and answer the questions.

When they place an insulator between the light bulb and the battery, they should find that there is no light coming from the light bulb.

They should notice that the light bulb intensity decreases as they add resistors to the circuit. Explain to them that the resistors are behaving like kinks in a hose. They are cutting off the flow of electrons like a kink in a hose cuts off the flow of water. The more resistors added to the circuit, the dimmer the light from the bulb will become.

They should also notice that as more resistors are added there is less heat in the wire.

❹ Remove the end of one wire from the battery and gently touch it with your finger to see if it is warm.

Record your results.

❺ Now place a piece of styrofoam or plastic in between one wire and the bulb.

Record your results.

❻ Remove the end of one wire from the battery and gently touch it with your finger to see if it is warm.

Record your results.

❼ Place a resistor between the light bulb and the battery on one wire. Observe any difference in the intensity of the light bulb.

Record your results in the chart.

Repeat with two or more resistors.

❽ Remove one end of the wire from the battery and gently touch it with your finger to see if it is warm.

Record your results.

Results

	wire only	wire + insulator	wire + resistor(s)
light bulb intensity			
temperature of wire			

Answer the following questions about your experiment:

▸ What happened when you connected the battery to the light bulb?
_____
_____

▸ What happened when you put a piece of styrofoam or plastic between the wire and the bulb?
_____
_____

▸ What happened when you put one or more resistors between the light bulb and the battery?
_____
_____

▸ What did the wire feel like to your fingers (with the wire only, with the insulator, and with the resistors)?
_____
_____
_____

Help the students draw and record some valid conclusions.

Conclusions

Review

(Answers may vary.)

Define the following:

static electricity   *charges that are not moving*

electric current   *charges that are moving*

voltage   *electric pressure*

resistor   *a material that slows down the flow of electric charges*

conductor   *a material that allows electric flow*

insulator   *a material that resists electric flow*

heat   *the transfer of heat energy from one object to another because of a difference in temperature*

Chapter 8: Magnets and Electromagnets

Overall Objectives	73
8.1 Magnets	73
8.2 Magnetic Fields	74
8.3 Electromagnets	75
8.4 Electromagnetic Induction	75
8.5 Summary	75
Experiment 8: Wrap It Up!	76
Review	81

Time Required

Text reading 30 minutes
Experimental 1 hour

Materials

metal rod (a large nail [such as an 8.9 cm [3½"] long 16d flathead nail or a screwdriver can be used for the rod)
electrical wire
10-20 paper clips
6v-12v battery
electrical tape
scissors
wire cutters
magnets, bar or horseshoe (2 or more)

Optional Materials

alligator clips (2)
thin magnet that can be cut (Section 8.1)
Section 8.2:
 iron filings
 60 ml (1/4 cup) corn syrup
 shallow dish
 iron nail

Overall Objectives

In this chapter the students will be introduced to magnets and their properties. Help the students understand that magnetic fields are produced in two ways:

- *by spinning electrons*
- *by moving charges, i.e., electric currents*

In magnets the electrons don't hop from one atom to another as is the case in electric current but, instead, spin on their axes. One important concept the students should understand is that electric currents create magnetic fields and magnetic fields can induce electric currents.

8.1 Magnets

While asking the students the following questions, it would be helpful for the students to have a few magnets to play with.

- *What happens if one magnet is brought close to another magnet?*
  (the other magnet will either be attracted or repelled)

- *What happens if you switch the direction of the first magnet (flip it over [horseshoe magnet] or use the opposite end [bar magnet])?*
  (the magnets will do the opposite from what they did in the first question)

- *What happens if you place your magnet near the refrigerator?*
  (it will be attracted to the refrigerator and will stick—if the refrigerator has steel doors)

- *Does your magnet stick to aluminum cans?*
  (No)

Have a discussion about the fact that magnets have poles. Poles are not "charged" like the rod and pieces of aluminum foil were in the Chapter 6 experiment, but a pair of poles is much like pairs of opposite charges. They are opposite fields that produce force.

Explain that in a permanent magnet, poles are caused by moving electrons, but in a magnet the electrons do not move from atom to atom. Instead, the electrons stay in one place and spin, much like a basketball that is spun on a finger. The basketball—or an electron—rotates around an axis and can spin in either direction.

All materials have spinning electrons, but not all materials are magnetic. Why? Explain that electrons in all materials spin in both directions. In materials where the atoms share an even number of electrons, the spins "cancel" each other out, and the material is not magnetic. However, in materials where there are an uneven number of electrons spinning, the extra spinning electrons can make the material magnetic when these electrons align themselves to create poles.

In Chapter 6 the students learned that like electric charges repel each other and unlike electric charges attract each other. The same rules of attraction apply to magnetic poles.

- *Like poles repel each other.*
- *Unlike poles attract each other.*

It is important for the students to understand that a major difference between electric charges and magnetic poles is that:

- *magnetic poles cannot be separated*

If a thin magnet is available that is easy to cut, it would be helpful to have the students try to separate the poles by cutting the magnet again and again until it is very small. Each little piece will still have two poles because the atoms themselves are little magnets.

8.2 Magnetic Fields

Have the students slowly bring one magnet close to another magnet. If the opposite poles of each magnet are brought close to each other, the students should experience a pull as the magnets approach but before they actually touch. Ask them why they think this might be happening.

Explain that a magnet produces a magnetic field that extends out from the magnet into the space around it, so a magnetic field affects the space surrounding the magnet. Even though the magnet physically ends at its edges, the magnetic fields extend beyond those physical ends.

There is an easy way to "see" the magnetic fields and, if time permits, the following simple demonstration should be performed using a magnet and some iron filings.

An easy way to collect iron filings is to place a magnet in a plastic bag and drag the plastic bag through some dirt. The iron filings will collect on the outside of the bag. Place the bag with the magnet still in it inside another bag. Then remove the magnet from the inner bag. The iron filings will release from the inner bag and collect inside the outer bag. Repeat this several times until a teaspoon or so of iron filings has been collected. Iron filings may also be purchased.

Mix the iron filings with 60 ml (1/4 cup) of corn syrup and pour the mixture into a clear shallow dish (the bottom portion of a clear glass butter dish works well). Place the dish on top of a regular-sized magnet. Have the students observe the iron filings and then wait for an hour and observe them again. The iron filings should align and look similar to the magnetic field lines shown in Section 8.2 of the textbook.

Explain to the students that the magnetic fields pass through the corn syrup without affecting the syrup, but because iron responds to magnetic fields, the iron filings line up with the magnetic field lines.

Magnets can make other objects temporarily magnetic. Using an iron nail, the students can induce a magnetic field in the nail by having the nail contact the magnet. When the magnet is removed, the iron nail will remain magnetic for some time.

8.3 Electromagnets

In the experiment for this chapter students will build a small electromagnet. Explain to the students that when an electric current flows through a wire, a magnetic field is created around the wire. When the wire is coiled, it can behave much like a bar magnet. The more electric current that passes through the wire, the stronger the electromagnet.

Explain that electromagnets can be quite strong and are often used in junk yards to lift heavy items such as cars. They are also convenient because they can be turned off.

8.4 Electromagnetic Induction

A magnet can also cause an electric current to flow through a coiled wire. In the experiment for this chapter the students will see that a coiled wire will create magnetic fields, but a magnet can also induce a voltage in a coiled wire. This is called electromagnetic induction. Faraday's Law describes electromagnetic induction in this way:

- *The induced voltage in a coil is proportional to the number of loops times the rate at which the magnetic field changes within those loops.*

This means that the more loops, the greater the voltage, and the faster the magnet is pulled back and forth through the loops, the greater the voltage.

8.5 Summary

Discuss the summary statements with the students.

Have the students read the experiment and then write an objective and a hypothesis. Some examples are:

Objective

- *In this experiment we will build an electromagnet.*
- *In this experiment we will explore the properties of electromagnets.*

Hypothesis

- *It won't matter how many coils we use, we won't be able to pick up more than a few paper clips.*
- *The more coils that are wrapped around the rod, the stronger the electromagnet.*
- *The number of paper clips we pick up will be proportional to the number of coils in the electromagnet.*

Have the students assemble the parts of the electromagnet. A large nail (such as an 8.9 cm [3½"] long 16d flathead nail) or a screwdriver works well as the metal rod. It should not be magnetized ahead of time.

Help the students trim the ends of the wire and fasten the ends to the battery. Alligator clips may be used instead of electrical tape. Different size batteries can be used, but a big 12v battery may work best with a screwdriver. Whenever the wire is connected to both the + and - terminals, the battery is running down, so the wire should be disconnected when the battery is not in use.

Before wrapping the wire around the rod, have the students touch the rod to a pile of paper clips and record the number of paper clips picked up by the rod.

Experiment 8: Wrap It Up!   Date: _____

Objective _____

Hypothesis _____

Materials

metal rod (a large nail or a screwdriver can be used)
electrical wire
10-20 paper clips
6v or larger battery
electrical tape
scissors
wire cutters

Experiment

❶ Cut the electrical wire so that it is .3-.6 meter (1-2 feet) long.

❷ Trim the plastic coating off the wire so that there is about 6 mm (1/4 inch) of exposed metal on each end of the wire.

❸ Tape one end of the wire to the positive (+) terminal of the battery. (Alligator clips may be used in place of tape.)

❹ Tape the other end of the wire to the negative (-) terminal of the battery.

❺ Take the metal rod and touch it to the paper clips. Record your results on the chart in the *Results* section.

❻ Coil the wire around the metal rod a few times. The wire must remain hooked to the battery.

❼ Touch the metal rod to the paper clips. Count the coils and record your results.

❽ Wrap another 1 to 5 coils around the metal rod.

❾ Touch the end of the metal rod to the paper clips. Record how many paper clips can be picked up.

❿ Continue adding coils to the metal rod and counting the number of paper clips that can be picked up. Record the results each time you increase the number of coils.

Have the students wrap the wire around the metal rod several times and then touch the metal rod to the pile of paper clips. Have them record the results in the *Results* section and then repeat these steps.

Each time more coils are added, have the students record the number of paper clips that are picked up.

Using a 12v battery, one student got the following results:

Number of Coils	Number of Paper Clips
10	0-very weak
15	1
20	2
25	8
30	12
35	15

Results may vary, but there should be an overall trend—as the number of coils increases, the number of paper clips the electromagnet can pick up also increases.

Results

Number of Coils	Number of Paper Clips

Graph your results.

Help the students graph their results.

The physics behind an electromagnet:

When a current flows down a wire, a magnetic field is created that rolls around the wire:

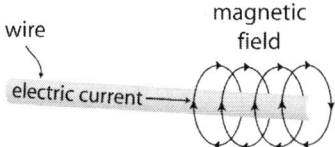

The strength of the magnetic field is proportional to the current; when there is more current, there is a stronger field. If you wind a wire into a coil, the fields from each part of the coil add up to create a net magnetic field that looks much like the field of a bar magnet:

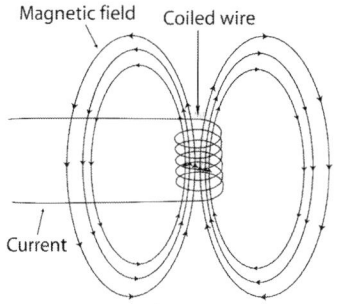

It can be shown that the strength of the magnetic field near the coil is proportional to the number of loops in the coil: more loops—stronger field. So the students should see a stronger magnet (one that picks up more paper clips) when:

- *the number of loops increases*
- *the current in the wire increases with a stronger battery*

Looking at the graph, have the students discuss their results and draw conclusions based on the data they have collected. Help them discuss any sources of error they might have encountered.

Conclusions

Review

(Answers may vary.)

What makes materials magnetic? <u>an uneven number of spinning electrons</u>

Why are some materials magnetic and others not? <u>only some materials have an uneven number of spinning electrons</u>

What are poles? <u>opposite ends of a magnet</u>

Opposite poles (**attract**/ repel). [Circle the correct word.]

Like poles (attract / **repel**). [Circle the correct word.]

What happens if an electric current flows around a metal rod?

<u>it induces a magnetic field</u>

What happens if a magnetic rod is pushed and pulled through a wire coil?

<u>it induces a voltage</u>

Draw a magnetic field.

Chapter 9: Light and Sound

Overall Objectives	83
9.1 Light	83
9.2 Waves	84
9.3 Visible Light	85
9.4 Sound Waves	86
9.5 Summary	86
Experiment 9: Bending Light and Circle Sounds	87
Review	93

Time Required

Text reading   30 minutes
Experimental   1 hour

Materials

- two prisms (glass or plastic)
- flashlight
- metal can, open at both ends
- aluminum foil
- rubber band
- laser pointer
- long wooden craft stick
- colored pencils
- duct tape (or other strong tape)

Optional Materials
(see Section 9.2)

- bowl filled with water
- styrofoam or other material that floats
- eyedropper

Overall Objectives

In this chapter students will be introduced to light and sound energy. They will look closely at waves and their properties, discuss the electromagnetic spectrum and visible light, and examine the difference between light waves and sound waves.

9.1 Light

Ask the students what they think light is. If a flashlight is available, have them explore some properties of light.

- *What happens to the light when you shine the flashlight in the dark?*
  (It shines in a straight line until it disappears or hits something.)

- *What happens to the light from the flashlight when you shine it during the daytime?*
  (It disappears immediately, and we can't see it.)

- *What happens if you put your hand in front of the light beam?*
  (The light is blocked.)

- *What happens if you shine the flashlight in a mirror?*
  (The light is reflected or "bounces back.")

- *What happens if you shine it on a non-mirrored surface?*
  (The light disappears into the surface.)

- *Can you "catch" the light in your hands?*
  (No)

- *Can you catch the light in a bottle or a box?*
  (No)

- *What happens to the light if you close your eyes?*
  (Nothing—we just can't see it.)

- *Can you hear light?*
  (No)

From these observations, the students should understand that light cannot be "captured," that it bounces back (reflects in a mirror), that it "disappears" in sunlight because it blends into other light, that it can be blocked (by their hands), that it gets absorbed by non-reflective surfaces, and that it cannot be detected by other senses such as touch or hearing.

Explain that light is actually a combination of electric and magnetic fields. Light is an electromagnetic wave. Therefore, because light is a wave, it will have the properties of a wave. (Light also has particle-like properties, but this concept is outside the scope of this level).

9.2 Waves

Discuss the nature of waves. Ask the students to describe what they know about waves, what waves look like, and how they move.

If time permits, have the students perform this short experiment.

- *Fill a bowl with water, then add a small piece of styrofoam or something else that will float. Using an eyedropper, let individual water droplets fall into the water. Observe how the water moves and how the small piece of floating foam moves.*

The students should see the water ripple, appear to bounce off the edges of the bowl, and then go back to the center. The small piece of styrofoam will more or less stay in one place. Explain that the water is mainly moving up and down and the disturbance from the water droplet moves outward.

Have the students look at the illustrations in Section 9.2 of the textbook and observe the parts of a wave. Point out that a wave has a peak and a valley, and the peaks are separated by a certain distance called a wavelength. The height of a peak from the center of the wave is called the amplitude.

Show them that the waves can be stretched or squeezed by moving the peaks farther apart or closer together. This changes the wavelength, but not the amplitude.

Explain that each color of light is an electromagnetic wave with a different wavelength. The students will look more closely at visible light in the next section.

Point out that we can only see a small part of the electromagnetic spectrum. Radio waves and microwaves are light—but we cannot see them.

Have the students look carefully at the textbook illustration of the electromagnetic spectrum, and point out that visible light is between infrared light and ultraviolet light. Show them that radio waves are much longer than visible light, and x-rays are shorter. Although we cannot see infrared light, we can often feel it. The heat from a campfire or from a glowing hot burner on a stove is mostly infrared light being absorbed by our skin and causing it to warm up. Likewise, we cannot see ultraviolet (UV) light, but it does have effects we can detect. UV light from the sun causes sunburns and suntans.

9.3 Visible Light

Have a discussion about the different colors of light. For example:

- *What are the different colors in a rainbow?*
  (red, orange, yellow, green, blue, violet)

- *Do you ever see yellow, purple, red and then green, in that order, in a rainbow?*
  (No)

- *Are the colors ever in a different order?*
  (No)

- *When does a rainbow occur?*
  (when it rains)

- *Do rainbows always occur when it rains?*
  (No)

Explain that each color of light is a wave at a different wavelength. Red is the same as violet except that the wavelength for red is longer. By squeezing a red wavelength or stretching a violet wavelength, each would become a different color.

Because colors have different wavelengths, when white light is split, the colors come out in a particular order. This is why the rainbow always goes from violet to red—red being the longest wavelength and violet being the shortest. A rainbow can be inverted (a double rainbow) but the colors will still be in the same order.

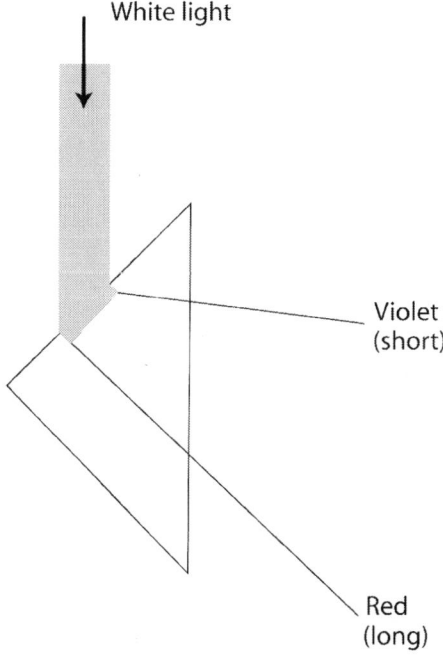

9.4 Sound Waves

Have the students think about sound. For example, ask:

- *What makes sounds?*
  (birds, musical instruments, anything that crashes, etc.)

- *If you place your ear on a table and have a friend tap the other end, what do you hear?*
  (a vibration)

- *When you speak, what makes the sound?*
  (vibrating vocal cords inside the throat)

- *What does your throat do as you speak?*
  (it vibrates)

Explain that sound waves are not the same as light waves. Sound waves are waves of air particles. When we hear sound, the particles in the air or in a table are vibrating in a wave and we pick up this vibration with our ears and hear sounds. Explain that there are different sounds just as there are different colors.

Frequency is the number of peaks in a wave that pass a given point in a given amount of time. Low frequencies have long wavelengths and high frequencies have short wavelengths. We call the frequency of a sound its pitch. High pitch means high frequency and short wavelength.

The intensity of a sound depends on its amplitude. Loud sounds have high amplitudes and soft sounds have low amplitudes. Discuss the meaning of decibel. Explain that the human ear can be damaged by loud noises. This is why workers in factories or in airports wear earplugs around loud machinery or near airplanes.

9.5 Summary

Discuss the summary statements with the students.

Experiment 9: Bending Light and Circle Sounds

Date: _____

Objective
_____
_____

Hypothesis
_____
_____

Materials

- two prisms (glass or plastic)
- flashlight
- metal can, open at both ends
- aluminum foil
- rubber band
- laser pointer
- long wooden craft stick
- colored pencils
- duct tape (or other strong tape)

Experiment

PART I: Bending Light

❶ Take one prism and shine the flashlight beam through it at the 90° bend. (See following illustration.) Have a wall or white board behind the prism.

Record your results in the *Results* section.

In this experiment students will examine some properties of light and sound.

Have the students read the entire experiment and then write an objective and a hypothesis. Some examples are:

Objective

- *We will examine light by using a prism to separate the colors.*
- *We will look at the wave nature of sound.*

Hypothesis

- *Sunlight shining through a prism will not separate into different colors.*
- *Sunlight shining through a prism will separate into different colors.*
- *The flashlight beam will not separate into different colors through the prism.*
- *The flashlight beam will separate into different colors through the prism.*
- *We will see waves with the soundscope.*
- *We will not see waves with the soundscope.*

Have the students shine the flashlight beam through the prism and then record their results. Have them take the prism outside and shine sunlight through it. Have them note any differences.

Getting everything positioned correctly can be tricky. Help the students angle the prism so that the light will pass through it. The resulting rainbow will be cast behind the prism.

Next, have them place two prisms together and shine the flashlight through both of them. Here they will observe a "double rainbow." The first rainbow will have red on the bottom and violet on top (or vice versa depending on how you shine the light through), and the second rainbow will be inverted—that is, exactly the opposite of the first rainbow. This can be hard to do, so you may need to help the students carefully adjust the angles and positions of the two prisms until the double rainbow is visible. Have them record their results.

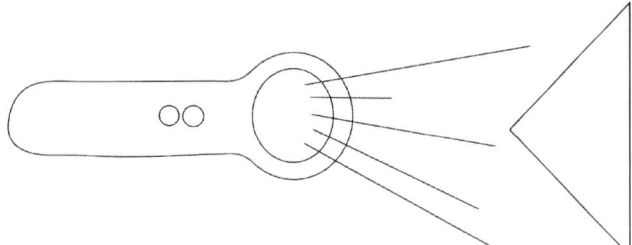

❷ Now take the prism and let sunlight shine through it from the same direction.

Record your results.

❸ Take the second prism and place it directly behind the first one, laying it flat on one of the short edges. Using the flashlight, shine light through the two prisms together. (See illustration.)

Record your results.

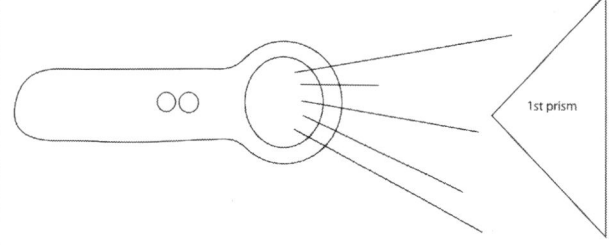 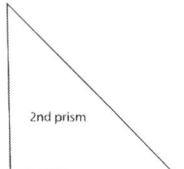

Results

PART I: Bending Light

What happens when you shine a flashlight through the prism?

_____
_____
_____

What happens when you put the prism in sunlight?

_____
_____
_____

What happens when you shine light through two prisms at the same time?

_____
_____
_____

Draw the shapes you see.

Have the students assemble the soundscope. Make sure the aluminum foil does not become wrinkled.

When working with a laser pointer, have the students be very careful not to shine the beam in their eyes!

The laser beam should reflect off the aluminum foil and onto an opposite wall. It will usually work best if the laser beam hits the center of the foil. Have the students record what they see as they speak into the can. Have them make both high-pitched and low-pitched sounds.

PART II: Circle Sounds

Assemble and experiment with a "soundscope."

❶ Take the metal can and make sure it is completely open on both ends.

❷ Place a piece of aluminum foil over one end of the can, and secure it with a rubber band. Be careful not to wrinkle the foil; try to keep it smooth.

❸ Fasten the craft stick securely to the metal can with strong tape.

❹ Place the laser pointer on the craft stick with the light facing the foil. It should look like the setup in the following diagram.

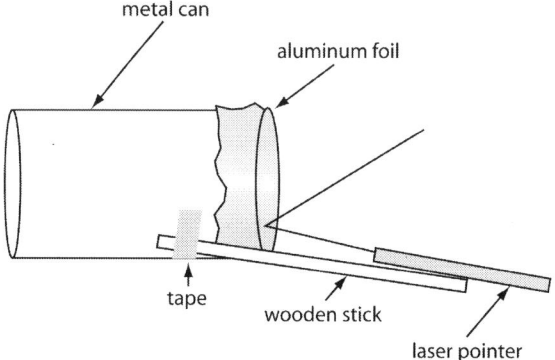

❺ Turn on the laser pointer. (Be careful not to point the laser directly into your eyes!) Observe the reflection on a wall or white board.

❻ Holding the can to your mouth, speak into it, and watch what happens to the reflected laser light. Record your results on the next page.

❼ Continue to speak or sing into the can, recording as many different shapes as you see.

CHAPTER 9 | 91
Light and Sound

Draw the shapes you see. *(Drawings may vary.)*

They should observe circles of different shapes. Explain to the students that these are waves. They look like circles because the laser pointer is stationary. If the laser pointer were moving, they would look like normal waves. The waves or circles show the vibrations of the aluminum foil, which are caused by the sound vibrations in the air. This experiment allows us to visualize sound.

Help the students draw conclusions from the data they collected.

Conclusions

PART I: Bending light

_____
_____
_____
_____
_____
_____
_____
_____
_____
_____
_____
_____

PART II: Circle sounds

_____
_____
_____
_____
_____
_____
_____
_____
_____
_____
_____

Review

(Answers may vary.)

Define the following terms:

electromagnetic wave   *magnetic and electric fields combined*

wavelength   *the distance between two peaks of a wave*

amplitude   *the height of a wave from the middle point to the top*

electromagnetic spectrum   *from radio waves to gamma rays*

visible light   *the part of the electromagnetic spectrum visible to the human eye*

pitch   *different wavelengths of sound*

frequency   *the number of peaks of a wave that pass a given point over a given amount of time*

Are radio waves sound?   *No. Radio waves are light waves that have long wavelengths.*

Chapter 10: Conservation of Energy

Overall Objectives	**95**
10.1 Introduction	95
10.2 Energy Is Conserved	95
10.3 Usable Energy	95
10.4 Energy Sources	96
10.5 Summary	97
Experiment 10: On Your Own	**98**
Review	**103**

Time Required

　　Text reading　　30 minutes
　　Experimental　　1 hour

Materials

　　student selected materials

CHAPTER 10
Conservation of Energy

Overall Objectives

In this chapter students will learn more about the connections between the various types of energy they have been studying, and they will learn about the law of conservation of energy. This law states that energy is neither created nor destroyed but is just converted from one form to another. When we hear about an energy crisis, this is referring to the depletion of usable energy, which is energy that can be converted into other forms of energy or work.

10.1 Introduction

Review the types of energy that have been studied, and discuss some ways in which energy is converted from one form to another. Some examples from the *Laboratory Workbook*:

- *The gravitational potential energy of a toy car is converted to kinetic energy as the car rolls down a ramp. (Experiment 3)*
- *The chemical energy in a battery is converted to light energy and heat energy. (Experiment 7)*
- *The chemical energy in a battery is converted to magnetic energy and mechanical energy. (Experiment 8)*
- *The electrical energy in a charged rod is converted to mechanical energy. (Experiment 6)*

10.2 Energy Is Conserved

Have the students look at the textbook illustration showing the toy car travelling down the ramp. Explain that when the toy car is sitting at the top of the ramp and is not moving, it has GPE (gravitational potential energy) but no KE (kinetic energy). Point out that as the car rolls down the ramp, it picks up KE but loses GPE. This is because, as it rolls down the ramp, the car loses height, and as it loses height, it loses GPE. Discuss the equation at the bottom of the illustration. At each point on the ramp, the total energy (which equals GPE + KE) stays the same. Explain that this is what is meant by conservation of energy.

10.3 Usable Energy

Strictly speaking, in the toy car example above, not all of the GPE is converted into KE. Because of friction, some amount of GPE is converted to heat instead. Explain that heat is "unusable" energy. That is, the heat cannot be converted into another form of energy. When we say that the energy is "lost," we really mean it is not useful anymore. Have the students think about other examples where usable energy is "lost." For example:

- *A light bulb converts electrical energy into light (useful), but also gets hot (produces heat energy which is not useful). In the end, the electrical energy and light energy are all "gone"—converted into heat energy.*

- *A car engine converts chemical energy in gasoline into useful mechanical energy, but eventually all of the mechanical energy gets dispersed as heat due to friction (from the air, engine, tires, and brakes). In the end, the gasoline is "gone" and the mechanical energy is "gone" too. All of the energy has been converted into heat.*

- *A battery-operated CD player converts electrical energy into music (useful), but also produces some heat (not useful). In the end, the electrical energy in the battery is "gone" and can no longer play the music.*

Eventually every kind of useful energy is converted to heat. This is an example of the Second Law of Thermodynamics in action. The energy is not really gone but has been converted into a useless form.

Explain to the students that when they hear the term energy crisis, it means that usable energy (energy that can do work by being converted into other forms of energy) is being used up.

10.4 Energy Sources

Have a discussion about different sources of energy. Explain that some sources of energy, like fossil fuels, cannot be renewed. That is, once they are used they cannot be replaced. Fossil fuels come from plants and animals that died a very long time ago. As the plant or animal decomposed, natural gas, oil, and coal were formed. Reservoirs of fossil fuels can be found in between rock layers under the ground. The fuel can be mined by digging or drilling into the ground and removing the fossil fuel. Ask the students to list some fossil fuels and how they are mined.

For example:

- *coal—dug from under the ground*
- *oil—holes are drilled into the ground and lined with pipes to remove the oil*
- *natural gas—found above oil and "mined" or brought to the surface with pipes*

Explain that most of the energy we use comes from fossil fuels. Discuss why some day the fossil fuels will no longer be available. Ask the students what we might do when the fossil fuels are gone.

Lead a discussion about renewable sources of energy, such as solar energy, wind energy, and energy that comes from water. Explain that some of these energy sources are expensive to use right now, but it may be possible in the future to use them with greater efficiency, and then they could be used as energy sources to replace fossil fuels.

10.5 Summary

Discuss the summary statements with the students.

For this chapter the students will design their own experiment. The goal is to use as many different forms of energy as they can and to convert the different energies into other forms of energy.

In the example given, the kinetic energy of a rolling marble is used to knock down a domino that has a cap filled with baking soda on top of it. The baking soda falls into vinegar and a chemical reaction is started.

The marble begins with GPE which gets converted to KE as it rolls down the ramp. The KE is used to convert the GPE of the elevated baking soda into KE as it falls. This releases the chemical potential energy (CPE) in the baking soda and vinegar, and a chemical reaction begins, producing CO_2 which then puts out the fire. The chemical energy is converted into heat energy and bubbles (gas).

Have the students think of ways this example might be extended. For example, the gas from the chemical reaction could be released into a small balloon or used to move a small piston.

First, have the students do several "thought experiments" by asking themselves how they might set up a series of small scenarios, like the example shown. Some of their ideas will not be practical, but have them use their imagination to think of different ways to convert energy.

Experiment 10: On Your Own   Date: _____

▸ This time you get to design your own experiment. The goal is to convert as many forms of energy as you can into other forms of energy.

Example:

A scenario can be designed in which energy is used to put out a fire. A marble is rolled down a ramp and bumps into a domino that has a small cap of baking soda on top of it. A chemical reaction is started when the baking soda falls into vinegar, which produces carbon dioxide gas that puts out the fire.

In this case, the rolling marble has kinetic energy which is used to convert gravitational potential energy into kinetic energy (the falling baking soda) which then starts a chemical reaction.

Using Energy to Put Out a Fire

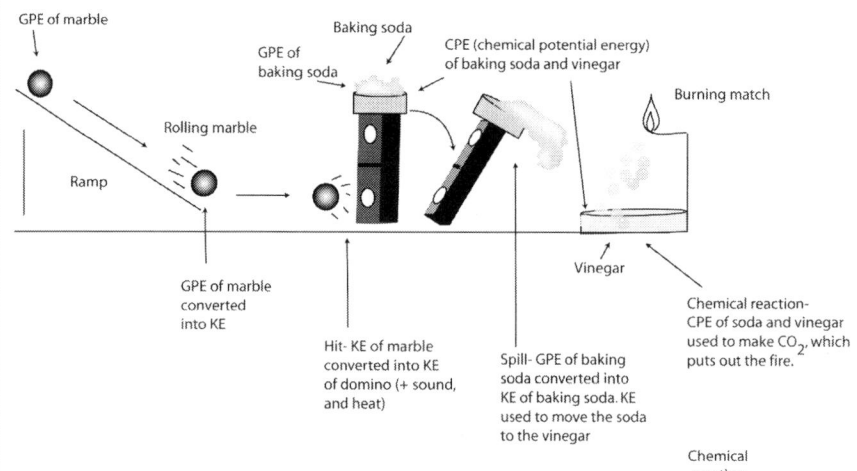

Use the following guide to design your experiment:

❶ Write down all the different forms of energy you can think of.

kinetic energy _____ _____

_____ _____

_____ _____

❷ Write down how these forms of energy can be represented.

kinetic energy

rolling marble	moving toy car	moving ball

CHAPTER 10 | 99
Conservation of Energy

Have the students think of different forms of energy and how they might represent these different forms. Help them narrow some of their ideas into practical applications.

| FOCUS ON MIDDLE SCHOOL PHYSICS
Teacher's Manual

Next, have them think about ways to link their different ideas. Have them start thinking about whether or not their ideas could work and have them start looking for items they might use in their experiment.

❸ Write down ways to connect two or more of these forms of energy and explain how one form will be converted into another.

moving toy car bumps into marble and starts it rolling

❹ Design an experiment to convert one form of energy into another. Give your experiment a title, and write an objective and a hypothesis. Write down the materials you will need, and then write down the steps you will take to collect the results. See how many different forms of energy you can convert. Make careful observations and draw conclusions based on what you observe.

CHAPTER 10 | 101
Conservation of Energy

Experiment 10: _____ **Date:** _____

Objective _____

_____

Hypothesis _____

_____

Materials

_____   _____   _____

_____   _____   _____

_____   _____   _____

Experiment

_____

_____

_____

_____

_____

_____

_____

_____

_____

Have the students design their own experiment and write an objective and hypothesis. Help them assemble a materials list, and then have them write down the steps of their experiment.

Have the students record their results whether or not their experiment worked. Have them write valid conclusions based on their results, and ask them what they might do differently next time.

Results

Conclusions

Review

(Some answers may vary.)

Describe the law of conservation of energy:

This law states that total energy is conserved. This means that energy is neither created nor destroyed.

What is the energy that is conserved? (Circle one.)

- kinetic energy

- potential energy

- (total energy)

- chemical energy

What is usable energy? *energy that can be converted into other forms of energy or used to do work*

Name one form of energy that is sometimes unusable.

heat energy or light energy